Our Magnetic Earth

Our Magnetic Earth

The Science of Geomagnetism

RONALD T. MERRILL

THE UNIVERSITY OF CHICAGO PRESS   CHICAGO AND LONDON

RONALD T. MERRILL is professor emeritus of earth and space sciences at the University of Washington. He is the coauthor of *The Magnetic Field of the Earth: Paleomagnetism, the Core, and the Deep Mantle* and recipient of the 2002 John Adam Fleming Medal.

The University of Chicago Press, Chicago 60637
The University of Chicago Press, Ltd., London
© 2010 by The University of Chicago
All rights reserved. Published 2010
Printed in the United States of America

19 18 17 16 15 14 13 12 11 10   1 2 3 4 5

ISBN-13: 978-0-226-52050-6 (cloth)
ISBN-10: 0-226-52050-1 (cloth)

Library of Congress Cataloging-in-Publication Data
Merrill, Ronald T.
  Our magnetic Earth: the science of geomagnetism / Ronald T. Merrill.
      p. cm.
  Includes bibliographical references and index.
  ISBN-13: 978-0-226-52050-6 (cloth: alk. paper)
  ISBN-10: 0-226-52050-1 (cloth: alk. paper) 1. Geomagnetism. 2. Paleomagnetism.
3. Dynamo theory (Cosmic physics).   I. Title.
  QC816.M474 2010
  538'.7—dc22   2010012964

Contents

Acknowledgments

I have received contributions to this book from numerous students and colleagues. Some, but not all of these, are Charles Barton, Mike Fuller, Erika Harnett, Mike McCarthy, Phil McFadden, Tom Quinn, Gerard Roe, and Sanjoy Som. Rob Coe and two anonymous reviewers read a draft of the entire book and made many valuable comments. My editor, Jennifer Howard, encouraged me throughout this project and made several helpful suggestions. Most of all, I thank my wife, Nancy, who read the text several times and insisted that I explain the material in a way she could understand it.

Magnetism and the Present Magnetic Field

If there had been a human with a compass 800,000 years ago, he would have found his compass needle pointed south rather than north. Twenty thousand years later Earth's magnetic field reversed.[1] The magnetic north and south poles swapped hemispheres. The compass needle also would have flipped. Reversals have occurred hundreds of times in Earth's history. Our magnetic field is constantly changing, even during our lifetimes. During the twentieth century, the strength of the field declined by 6 percent, and the magnetic north pole moved northward by nearly $10°$ in latitude.

What causes our magnetic field to be so restless? What are the consequences of this restlessness? Do variations in Earth's magnetic field relate to magnetic storms that sometimes cause major power outages? Will they affect our climate? Will animals that sense, and use, the magnetic field encounter serious problems? These are just a few of the questions I have been asked by students and professional scientists. As a professor specializing in geomagnetism (the study of Earth's magnetic field) at the University of Washington, the questioners thought I would always be able to provide definitive answers. Sometimes I could and sometimes not.

In this book, I cover a broad range of topics dealing with magnetism in the earth and planetary sciences. I also provide you with the reasons why geomagnetists arrive at their conclusions. I will do this without using the mathematical language that we geomagnetists often use. To make our journey in geomagnetism more alive, I will provide some historical context and introduce you to some of the characters I have met during my career.

In this chapter, magnetism and magnetic fields are treated first, as they are at the foundations of geomagnetism. Then our knowledge of the present field is developed in a historical context.

Our bewilderment and fascination with magnetism often began when we were introduced to magnets as children. We found it strange that two bar magnets in one configuration resisted being pushed together, while in the opposite one they attracted each other. Why? In my case, I felt the need to consult someone of authority. I turned to my father, who as a parent surely must know the correct answer. I was fortunate, because as an authority goes, he had the correct credentials—a PhD in theoretical physics. He explained to me that there are two ends to a magnet, a plus (or north) end and a minus (or south) end. When the two ends facing each other have the same sign, they repel each other. When they have opposite signs, they attract. He repeated this explanation a few different times to make sure I understood. Many of you will have received a similar explanation from an "authority," perhaps a high school science teacher. What was your reaction to your authority's explanation? I remember my reaction to my father's explanation: I must be dumb. I could see the magnets were attracted to each other in one alignment and were repelled in the opposite one, but I wanted to know why. I thought my authority must have given me the correct explanation, but I was simply not smart enough to understand it. I will say more about his answer and authority arguments as we travel through this book.

History illustrates that I was not alone in being puzzled by the seemingly mysterious properties of magnets.[2] I suspect that humans have been interested in the origin of magnetism ever since it became recognized that some minerals were magnetic. Magnetism was investigated over 2,000 years ago in two separate political systems, one centered on the Mediterranean Sea and one centered in China. These early studies focused on lodestone, which is more commonly known today as the mineral magnetite, an iron oxide (Fe_3O_4). It was mined in the province of Magnesia (Thessaly, Greece), and it became known by the Latin word *magneta*, from which the modern term "magnet" derives.

William Gilbert of Colchester (1544–1603), often referred to as the father of magnetism for his reliance on the (now-called) experimental method, attributed the attractive powers of lodestone to it having a soul, similar to the views of the Greek animist Anaxagoras around 460 BC. Gilbert's most important legacy came with the publication of *De Magnete* in 1600. In it, he reproduced many of the results of Petrus Peregrinus, whose remarkable experiments with spheres of lodestone in 1269 emphasized the dipolar nature of magnets: they always come with two poles, a plus one and a minus one. Gilbert recognized that Peregrinus's spheres could be viewed as models of Earth. Gilbert pointed out that "rays of magnetick virtue" spread out in every direction in an "orbe." The center of this orb is not at the pole, as had been conjectured by Peregrinus and others, but in the center of the stone. Building on this, Gilbert was the first to conclude that Earth itself behaves like a huge magnet. Gilbert also noticed that Earth's magnetic and geographic poles were in close proximity, arguing that this showed Earth's rotation about its axis stemmed from its magnetism. Later the philosopher René Descartes (1596–1650) exorcised the soul from lodestone and embraced Gilbert's view that Earth could be viewed as a large magnet.

Today we believe the main magnetic field originates from electric currents flowing in Earth's iron-rich core. (The process by which these currents are generated is the so-called dynamo, a concept that will be discussed in more detail later.) Gilbert and others could not have arrived at such a conclusion in their time because scientists were not aware of electric currents until the studies of Stephen Gray (1666–1736). The connection between electric currents and magnetism was not recognized until much later by Michael Faraday (1791–1867) and Hans Christian Oersted (1777–1851). In Gilbert's day, scientists were unable to construct the theory we accept today for the origin of Earth's magnetic field because the crucial physics had not yet been discovered. They had little recourse but to attribute the origin of the magnetic field to permanent magnetism. It makes one pause and wonder what incorrect explanations we hold today because we have not yet developed the crucial underlying science.

*   *   *

Even today it is sometimes difficult to know whether a magnetic field has arisen from electric currents or from magnetization, as I learned while serving on NASA's Lunar and Planetary Science Review Panel from 1978 through 1980. Well before the manned (Apollo) missions brought back lunar samples, potential astronauts had been trained in the southwestern United States to map rocks in the same way geologists did on Earth. At that time geologists often used a pace and compass technique: pacing to estimate distance and a magnetic compass to estimate direction. This training went on for several months until someone pointed out that the Moon had no magnetic field, which makes it impossible to rely on a compass for direction. When the first lunar samples were later returned to Earth, scientists were shocked to find some of them were magnetized. How could they be magnetized without the Moon having a magnetic field? Some scientists thought an error must have occurred. NASA even arranged to have a lunar sample sent back to the Moon and returned to Earth to make sure the magnetization was not accidentally acquired during transit. Over time, further missions showed that many lunar rocks were magnetized, and magnetic field measurements from various spacecraft showed that the Moon has a weak magnetic field stemming from a magnetized crust. The lunar magnetic field is complex, exhibiting many magnetic poles (where the field is vertical). Because the lunar magnetic field is weak and complex, magnetic compasses would be essentially useless on the Moon. It remains a mystery as to how the lunar crust became magnetized. Our best guess, and it really is little more than a guess, is the Moon had a dynamo-produced field for its first half billion years or so, as does Earth today (chap. 3).

We can gain more insight into magnetic fields by considering a different field, that of gravity. Like the magnetic force, the gravitational force acts at a distance. When I was a child, it did not seem as strange to me as did the magnetic force. Perhaps this is because we are constantly experiencing the gravitational force and even measuring it—one's weight is a measure of the gravitational force. A force is a push or pull specified by a direction and a magnitude. The gravitational force exerted by Earth is downward toward Earth's center, not sidewise or upward. The magnitude of the gravitational force between two objects is proportional to the product of their masses divided by the square of the distance. If the distance between two objects is doubled, the force decreases by a factor of four. After many measurements were made to confirm this, scientists accepted that they had a "law of gravity," first mathematically formulated by Sir Isaac Newton in 1687. Although tests of this law indicate that it is an excellent one, we have not answered the question why does this law exist? We simply have observed that it does, and like our best laws, Newton constructed a mathematical relation allowing us to make accurate predictions.

My wife and I weigh different amounts. That is, the gravitational force of Earth acting on my wife is different from that acting on me. Yet we would like to describe Earth's gravity at a particular location by a single number. This would allow us to do many things, including making comparisons of Earth's gravity to that of other planets. Earth's gravity is, for example, different from the Moon's gravity; we weigh about one-sixth as much on the Moon as on Earth. The concept of a gravitational field allows us to do this. One can determine the gravitational field of Earth by dividing the gravitational force by the mass it attracts. In other words, the gravitational field is the gravitational force per unit mass. This means Earth's gravitational field depends only on the mass of Earth and not the mass of the object Earth is attracting. If we want to know the gravitational force acting on, say, my wife, we can obtain this by multiplying her mass by the gravitational field.

A similar definition can be used for the electric force between two electric charges, which is proportional to the product of the charges divided by the square of the distance between them. The electric field is defined as the electric force on a unit of positive charge (such as that of a proton). The electric field does not depend on the magnitude of the charge that the electric field acts upon.

Scientists define a magnetic field in an analogous way: it is the force on a unit of positive magnetic charge. But there is a catch. No one has found an isolated magnetic charge, a magnetic monopole. Scientists have never even found indirect evidence that a single "magnetic charge" exists, unlike that in electricity where it is well documented that an electron has a single negative charge (an electric monopole) and a proton has a single positive charge. Nevertheless, the Nobel Prize–winning physicist Paul Dirac (1902–1984) postulated the existence of magnetic monopoles based on symmetry arguments. In particular, Dirac demonstrated that the famous Maxwell equations could be modified to include magnetic monopoles.

The equations modified by Dirac were four partial differential equations developed by James Clerk Maxwell (1831–1879) that unified electricity and magnetism. Rather than speaking of separate electric and magnetic fields, one could speak of an electromagnetic field. When electric charges move to produce an electric current, a magnetic field is simultaneously produced. Maxwell even used these equations to show that light traveled as an electromagnetic wave. More than one advanced course (graduate-student level) in the twenty-first century began by the professor writing the Maxwell equations down on the board and proclaiming that is all there is to electricity and magnetism—only the details require explaining, which will take the remainder of the year. While an exaggeration, such a statement illustrates the central role that the Maxwell equations play in electricity and magnetism.

Dirac showed that a simple modification to Maxwell's equations produced a symmetry between electricity and magnetism. Just as electric currents produce magnetic fields, the modified Maxwell equations permit magnetic currents, which produce electric fields. Although symmetry arguments are powerful instruments for advancing physics, so far no one has demonstrated that a magnetic monopole exists. But scientists are still searching for one!

You might wonder, if there are no known monopoles, what produces the magnetic field of magnets? Consider a simple bar magnet—a magnet with a length much greater than its width. Suppose we decide to cut this magnet into two parts in an attempt to separate the plus end from the minus end. This would result in two magnets, each with its own plus and minus ends.[3] In principle, we could repeat this experiment as often as we like. But if we did, we would always obtain magnets with two poles, a plus and a minus one. We could continue this thought experiment until we reduced the magnet's size down to a single electron. The electron is a fundamental particle, and further subdivision appears impossible. The electron still has two poles and behaves like a magnet. We refer to this magnet as a "point dipole," or simply "dipole." The word "dipole" refers to the fact that there are two poles—a plus one and a minus one. It is sufficient for us to accept as an observational fact that there is a magnetic dipole field, which is experimentally found to be tied to the electron's spin. Just as a moving electron produces a magnetic field, so does a rotating (spinning) electron.

Unlike a monopole field (for example, the electric field from an electron), the magnitude of a dipole field falls off as the cube of the distance from the source. If the distance between the point where the field is measured and the dipole's position is doubled, the magnitude of the field is reduced by a factor of 1/8. Because a dipole consists of plus and negative magnetic charges, it both repels and attracts any other magnetic charge. Because of these opposing effects, the magnetic field from a dipole falls off faster with distance than it would if only one magnetic charge were present.

The sources of Earth's magnetic field cannot be uniquely determined from measurements made at or above Earth's surface, a point I will return to later in this chapter. Although this causes us some difficulties in explaining the origin of Earth's magnetic field (as we shall see in chapter 3),

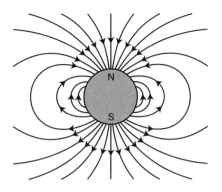

FIGURE I.I The geocentric axial dipole field is characterized by field lines entering Earth at the north pole and exiting at the south pole. The field is horizontal at the equator. The original figure from the United States Geological Survey was redrawn by Beth Tully.

it allows us to represent much of Earth's magnetic field by the geocentric axial dipole field shown in figure 1.1. This dipole (which can be visualized as a bar magnet) is at Earth's center and aligned along Earth's rotation axis. We represent the field this way, even though the actual magnetic field does not originate from a strong dipole at Earth's center. We do this because it is mathematically convenient and because it is easy to visualize. This geocentric axial dipole field is vertical at two points: the geomagnetic north and south poles. These two points coincide with the geographic poles because the field is along the rotation axis. The field lines are also shown. While these lines are not real physical entities, they are useful devices to describe the direction and magnitude of the magnetic field at any locality.

Consider a magnet suspended from a string. It would be a compass if it were constrained to rotate only in a horizontal plane. Then it would point to the north pole. However, if we allow the suspended magnet to rotate freely, its north end would dip downward in the Northern Hemisphere and upward in the Southern Hemisphere. It would only be horizontal at the equator. It would lie along the field lines shown in figure 1.1, which emerge from the Southern Hemisphere and reenter Earth in the Northern Hemisphere.[4] The magnetic field is said to have a downward dip in the Northern Hemisphere and an upward dip in the Southern Hemisphere. This dip, also referred to as the magnetic inclination, is measured positive in the downward direction. It is +90° at the north pole and −90° at the south pole. The magnetic field at the surface is twice as large at the poles, where the field lines are closer together, than it is at the equator, where the field lines are further apart. Earth's axial dipole field has a strength near 60,000 nT (nanotesla, or a billionth of a tesla) at its poles and half that at the equator. (The field is also commonly given in centimeter-gram-second units at the poles as 0.6 gauss.) Earth scientists often refer to the field strength of Earth as the magnitude of the field at the poles, while space scientists often refer to the field magnitude at the equator. This can create some misunderstanding, which can be compounded because even scientists sometimes get confused on the units used in electromagnetism. I will usually avoid using units in this book.

*   *   *

Earth's magnetic field has an upward inclination in Australia, where I had traveled to attend an international meeting in earth sciences. Although I had planned not to think about magnetic fields or magnetism during the first day after arrival, within a few hours in the state of Queensland I was discussing them.

"Magnetic fields are common on Earth and throughout the universe," I said. Nodding in agreement with me, the manager of the shop in Brisbane replied, "Everything is magnetic." In an attempt to overcome jet lag, I had taken a long run, and on the way back to my hotel, I had stumbled on an entire store devoted to magnetic therapy. I knew that everything containing electrons, which is practically everything, is magnetic. Electrons have spin and a dipole magnetic field. When electrons move, they produce an electric current and an associated magnetic field. When Joseph J. Thomson (1856–1940) discovered the electron, for which he received a Nobel Prize in 1906, electric current had already been defined as the flow of positive charges. The flow of electrons in one direction is equivalent to the flow of positive charges in the opposite direction. Experiments indicate that an electric current produces a magnetic field, as shown in figure 1.2.

My agreement with the manager was about to end as she tried to explain to me how magnetic bracelets work. She explained, "Hemoglobin contains iron and is therefore magnetic. The magnetic bracelet attracts the blood to the person's wrist and reduces arthritis." I decided to avoid using an authority argument. I didn't tell her I was a professor specializing in geomagnetism and was in Brisbane at an international scientific meeting to deliver an invited talk on the origin of magnetism in small magnetic particles. Instead I asked, "Why, then, doesn't the blood just pool up around the wrist?" She looked impatient. I moved the bracelet a couple of centimeters (slightly less than an inch) above a metal paper clip. The paper clip didn't budge. "Pretty weak magnet, isn't it? Are you sure you are not

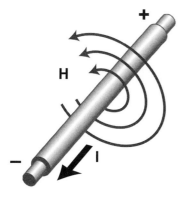

FIGURE 1.2 When an electric current, *I*, flows in a wire, it produces a magnetic field, *H*, that curls around the wire. The original figure from *Wikimedia Commons* was redrawn by Beth Tully.

dealing with a placebo effect?" By this time the manager had decided she was dealing with a troublemaker. I hastily departed the shop in the direction of my hotel.

I am certain I could have induced a stronger magnetization in the paper clip had I moved the magnetic bracelet closer to it. It would have exhibited paramagnetic behavior,[5] which means it would have been attracted to the magnet even though it was not permanently magnetized. Other materials, such as the mineral halite—which is known to us as common table salt (NaCl)—exhibit diamagnetism. Diamagnetic materials are repelled by magnetic fields. Diamagnetism and paramagnetism are examples of induced magnetizations. They only occur in a material in the presence of an applied magnetic field (including Earth's magnetic field). Remove the magnetic field, and the induced magnetization vanishes. "Magnetic susceptibility" is used by scientists to describe how strong the induced magnetism is: the greater the susceptibility, the greater the induced magnetization. Table salt has a tiny negative susceptibility. It has a small induced magnetization pointing in the opposite direction of the external magnetic field. It is so small that scientists can only measure it by using sensitive instruments.

In 1831 Michael Faraday captured the essence of diamagnetism in a general mathematical law. Heinrich Lenz (1804–1864) developed a more specialized version in 1834: electric currents are induced in a material, producing an internally generated magnetic field that opposes any applied magnetic field. We now know that those currents are due to modifications in electrons' orbits, which produce a net magnetic field opposing the applied field. In contrast, paramagnetism requires that an atomic dipole moment exists. An atomic moment refers to the fact that some atoms look like they possess a tiny bar magnet or dipole when viewed from distances that are great with respect to the size of the atom. This atomic moment occurs when the (vector) sum of all the individual dipole moments associated with the spins of electrons does not vanish. Paramagnetism is a result of atomic dipole moments (magnets) exhibiting a slight alignment along the direction of the applied magnetic field. (Lenz's law still operates in a paramagnetic material. However, the diamagnetic effect is smaller than the paramagnetic one.)

Hemoglobin carrying oxygen in arteries is diamagnetic. Contrary to the Brisbane shop manager's contention, oxygenated blood is slightly repelled by a magnetic field. Although magnetic therapy is a multimillion-dollar-a-year business, it does not have a strong scientific foundation.[6]

Another diamagnetic material is water, which is the major constituent of many animals, including frogs. In 1998 Andrey Geim and Jan Maan levitated a frog in a strong magnetic field. This is not an antigravity experiment. It simply illustrates that gravity is a relatively weak force. One can use the diamagnetism of water (and other materials) in a frog to counter the entire gravitational field of Earth. The frog was repelled by a strong magnet pole beneath it and attracted by a strong one above it. Geim and Maan also levitated many other substances, including pizza. For their efforts, they were awarded the Ig Nobel Prize, by the journal *Annals of Improbable Research* and the Harvard Computer Society. This prize is awarded for research that "cannot or should not be reproduced."

*   *   *

Alongside induced magnetism, there is permanent magnetism. While it can only be completely understood using quantum mechanics, it is critical to our understanding of geomagnetism. I'll hit the highlights, using classical physics.

A theorem was proved during the first half of the previous century showing that magnetism, including induced magnetism, could not exist if classical physics, such as exemplified by Maxwell's equations, accurately described all of nature. For example, the theorem proves that the magnetic fields associated with electron orbits producing diamagnetism in a grain of table salt would average to zero. For every electron orbiting in a clockwise direction, another would be orbiting in an anti-clockwise direction. The net magnetic fields associated with these electron orbits would vanish: no diamagnetism would occur. However, when quantum mechanics is applied, the average magnetic field associated with electron orbits in table salt no longer vanishes and diamagnetism occurs. Not all electricity and magnetism can be explained by Maxwell's equations.

Albert Einstein (1879–1955) was a pioneer of what used to be called the old quantum mechanics. He received the Nobel Prize in 1921 for mathematically explaining how a quantum of light had to be absorbed before a photoelectric material could emit an electron. Eventually it became known that an electron (and light) could be described as both a particle and a wave, an oddity introduced by quantum mechanics. The electrons emitted from photoelectric materials have discrete energies. Contrary to the classical theory of Maxwell, energy is not infinitely divisible. Erwin Schrödinger (1887–1961) ushered in the new quantum mechanics in 1926 by publishing his famous wave equation. A particle, such as an electron, is described by a wave function, a solution to the Schrödinger equation. This wave function can be used to determine the probability that an electron is at any particular location. John Van Vleck (1899–1980), who received the Nobel Prize in Physics in 1977, is often considered the father of the quantum mechanical theory of magnetism, because he was the first to use quantum mechanics to explain paramagnetism.

My father was familiar with all this: he was Van Vleck's first PhD student. Yet when I was a boy, I heard my father explain the origin of magnetism to adults using the same words he used to explain it to me. I wondered at the time whether these adults got it or whether they felt dumb like I did. I now recognize my father's explanation was a very good one. It is a law that can easily be tested, even by little boys. The question "Why is this law valid?" ultimately has no answer. Unlike mathematics, which uses deductive reasoning, science is based on inductive reasoning. A law is established when the observational evidence is so convincing that the probability of it not being observed can be dismissed. But what is the probability level required to do this? Consider the situation when the first egg in a carton is contaminated with salmonella. Would you be concerned the next egg might be similarly contaminated? If you found the first eleven eggs in the carton were contaminated, would this then convince you the last egg is? You have not proved the last egg is contaminated, but the evidence this is so mounts with each additional egg sampled. At what point is the evidence sufficient to make a contaminated egg law? The same type of inductive reasoning applies to scientific laws. The probability level required to establish some scientific finding depends on the degree of conservatism one has. Some scientists are so conservative they will only describe the way in which they conducted an experiment and the numbers they obtained in that experiment. In contrast, others will take giant intuitive leaps based on scant evidence. We will encounter examples of these extremes later in this book.

All physics requires the use of a "law" established by experiments, regardless of whether one uses classical or quantum physics. I will not offer quantum mechanic explanations for magnetism here, even though they are technically preferred. Instead, I will use a more classical approach to provide insight into permanent magnetization.

Let's consider materials at absolute zero temperature ($-273\,^{\circ}\mathrm{C}$), where one does not have to worry about the effects of heat, which can affect magnetism. Some materials contain atoms and molecules with an atomic dipole moment and some do not. The (vector) sum of the individual dipoles associated with the spins of electrons in table salt vanishes: there are as many spins pointing in one direction as there are in the opposite direction. A molecule of table salt, NaCl, has no net dipole moment (bar magnet) and hence cannot exhibit permanent magnetism. This is not the case for iron, which exhibits atomic moments on its atoms. Even in the absence of an applied magnetic field, these tiny magnets line up parallel to each other, an example of ferromagnetic ordering. It is the collective ordering of these atomic moments that produces the relatively strong magnetic field of an iron magnet.

The notion of magnetic order is of central importance in magnetism. A magnetic disordered state, in which the atomic moments are in random directions, exhibits no magnetism. Ferromagnetic order is shown in the top of figure 1.3. Materials with such ordering, such as iron, exhibit ferromagnetism at low temperatures. As is explained at the beginning of the next chapter, ferromagnetic order decreases with increasing temperature and vanishes at the so-called Curie temperature. Above the Curie temperature, the material is paramagnetic: it can acquire an induced magnetization, but it cannot exhibit a permanent magnetization.

Quantum mechanics also permits other ways the atomic magnets can be ordered, as first shown by Louis Néel (1904–2000), for which he received the Nobel Prize in Physics in 1970. Néel hypothesized that some materials could have adjacent magnet moments that alternate by 180° in direction, such as shown in figure 1.3 (the middle figure). Such materials exhibit antiferromagnetic order. Néel's prediction has been confirmed: an example of an antiferromagnetic mineral is ilmenite ($FeTiO_3$). Néel hy-

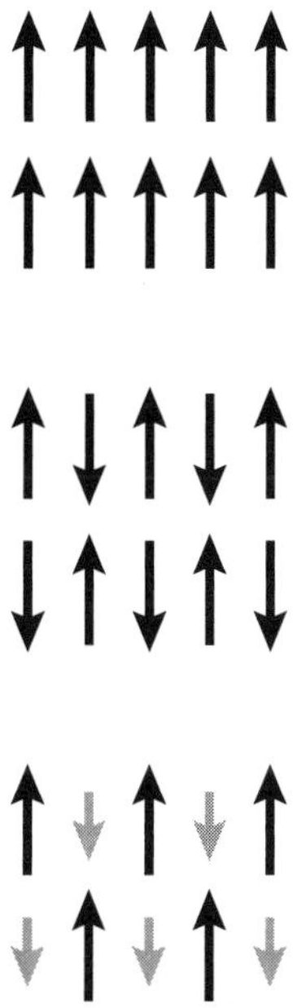

↑↑↑↑↑

↑↑↑↑↑

↑↓↑↓↑

↓↑↓↑↓

↑↓↑↓↑

↓↑↓↑↓

FIGURE 1.3 The upper figure illustrates ferromagnetism, in which the adjacent atomic moments are pointing in the same direction. The middle figure illustrates antiferromagnetism, in which adjacent atomic dipoles are antiparallel to one another. The lower figure illustrates ferrimagnetism, in which the downward antiparallel atomic dipoles are less strong than the upward-pointing ones. Figure drawn by Beth Tully.

pothesized yet another structure involving antiparallel coupling, as shown in the bottom of figure 1.3. Ferrimagnetism resembles antiferromagnetism, but the adjacent atomic magnets have different strengths. This also has been confirmed: an example of a ferrimagnetic mineral is magnetite (Fe_3O_4). (Magnetite has two different electrically charged atoms of iron: a ferrous ion that has one less electron than a ferric ion.)

Antiferromagnetism is not magnetic in the usual sense. It does not, for example, attract iron to it, because the magnetic fields associated with individual atoms cancel each other out. Ferromagnetic and ferrimagnetic materials are magnetic in the usual sense. The magnetic structures of all these materials originate from a force (the exchange force) that comes from quantum mechanical considerations.[7]

You may be taken aback by the above explanation of magnetism: all I have essentially said is "It exists and you should accept it." But isn't this explanation similar to the one you have for gravity? I suspect many of you accept the explanation provided earlier involving Newton's law of gravity. But why is this law true? Some of you may know that Newton's law has been superseded by a law of gravity developed by Einstein, more commonly know as the general theory of relativity. Even well-established laws evolve with time. Newton's law of gravity implicitly assumes mass is not a function of speed. Yet according to Einstein's theory of relativity, the mass of any object increases as its speed approaches that of light. This was subsequently confirmed, much to the relief of Einstein. The law is now well accepted. Yet it, too, will likely evolve with time, because it is now appreciated that the general theory of relativity and quantum mechanics are incompatible in some instances, such as in the vicinity of a black hole. (A black hole refers to an astronomical object with a gravitational field so large that not even light can escape from it.)

* * *

Electrons are not the only particles that have a spin and associated magnetic field. Protons and neutrons also do. However, the magnetic moment (strength) of the spin of an individual proton or neutron (bar magnet) is only about 1 percent of the electron's magnetic moment. The magnetic field of nuclei can essentially be ignored in our discussions of magnetism. Nevertheless, the nuclear magnetic moment does provide valuable information on the nature of materials. For example, medical technicians image the soft tissue in our bodies, such as cartilage, using MRI (magnetic resonance imaging). MRI measures effects associated with the magnetic fields of nuclei, particularly those of hydrogen atoms (each of which has a nucleus consisting of a single proton) in water. In contrast, X-rays are more useful for imaging hard parts of our bodies, such as bones.

The foundations of nuclear magnetic resonance (NMR) were independently developed by Felix Bloch (1905–1983) and Edward Purcell (1912–1997), who received the Nobel Prize in Physics in 1952 for their efforts. To understand resonance, consider a swing in a playground. The swing has a natural frequency at which it moves back and forth. If you continue to push the swing at its highest point, you amplify the signal. The swing goes higher. You have supplied energy to the swing, which it absorbed to increase its amplitude (peak height). The amplification occurs because the frequency you pushed it is the natural frequency of the swing: it resonates. I will not describe in detail how nuclear magnetic resonance works, but it involves the absorption of energy by a material from an alternating magnetic field applied perpendicular to a very strong magnetic field.[8] Bloch and Purcell adjusted the frequency of the alternating field until resonance occurred and energy was absorbed. Later Richard Ernst received the Nobel Prize in chemistry in 1991 for developing a way to rapidly carry out NMR analyses. His method involved using different radio frequencies at the same time to find the resonances.

Although the history of magnetic resonance involves many more Nobel Prizes, I will cut this story short by only mentioning some of the most recently honored work using NMR in medicine. Raymond Damadian published an important paper in 1971 showing that NMR could be used to distinguish cancerous tissue from normal tissue in vitro (in a test tube). He received various honors for this work, and he also patented his technique, the first patent for NMR. However, when the 2003 Nobel Prize in Medicine was awarded to individuals who developed MRI, Damadian was not included in the prize. (The word *nuclear* is not used in medical work, apparently because it has a bad connotation associated with harmful nuclear radiation, even though such radiation is not involved.) The prize went to Paul Lauterbur and Peter Mansfield. Damadian was irate. One can award the prize to three people, the maximum number allowed. He was convinced that he deserved a share of the prize. He argued that the Nobel Committee should rectify their mistake of not including him. He made a public show of discontent. Eventually a series of full-page advertisements appeared in the *New York Times*, the *Washington Post*, and the *Los Angeles Times*, requesting readers to write to the Nobel Committee to rectify this error. The heading on the October 30, 2003, advertisement in the *New York Times* read: "This Year's Nobel Prize in Medicine. 30 years of proof that this shameful wrong must be righted."

I believe it is not fruitful to second-guess why certain people receive prizes while others don't. Naturally, mistakes are sometimes made, as evidenced by Egas Moniz being awarded the Nobel Prize in 1949 for the use of lobotomy to treat psychoses. However, scientists familiar with MRI point out that Damadian's technique could not be used in practice to find cancer. His patent suggested NMR be used to scan the human body, but it did not provide any description of how this might be done. In contrast, Lauterbur and Mansfield developed a technique to use MRI to image the soft tissue in our bodies. Even Damadian recognized they deserved a share of the Nobel Prize.

* * *

No one knew about resonance half a millennium ago. For that matter, no one knew about electric currents or about the nature of atoms. Quantum mechanical explanations of magnetism were not possible then. Although scholars did know of magnetic minerals, the origin of the mineral's magnetism was controversial. Perhaps some materials were magnetic because they had a soul.

Scholars also knew Earth had a magnetic field and that it was useful for navigation. The Chinese had developed the first magnetic sea compass sometime between 850 and 1050 AD.[9] Nevertheless, there were several navigation puzzles that remained unsolved even at the end of the Middle Ages.

When one gives the coordinates of some location on Earth's surface, it usually is in terms of (meridians of) longitude and (parallels of) latitude. Any particular longitude is described by a great circle through the poles. Latitudes are circles centered on the rotation axis, including the 0° latitude, which is a great circle called the equator. Mariners used the stars to determine the longitude and latitude at sea very effectively, except on cloudy nights when the use of a magnetic compass became their primary method for navigation. Most of the early compass needles, made of magnetite, floated in a bowl of water and often needed to be remagnetized.[10]

Little understanding of the complexity of Earth's magnetic field existed before the nineteenth century. However, there was widespread acceptance of the benefit of accurate navigation, because this would prove of immense value in trade and war. A written decree by Pope Alexander VI in 1493 divided the world into two hemispheres and gave Spain the rights of trade in one hemisphere and Portugal in the other. This division assumed there was a prime meridian dividing Earth's surface into two hemispheres. Finding this prime meridian became crucial to determining the rights of those maritime powers. Because of the advantages that accurate navigation would bring to a country, several maritime powers promoted the search for this or some other navigational Holy Grail. In 1567 Spain's King Philip II, and thirty-one years later his successor King Philip III, issued large monetary rewards for the solution to the magnetic navigation problem. Similarly, large rewards were later offered by other maritime powers, such as Holland and England.[11]

To appreciate the problems encountered by the ancient mariners, we need to increase our understanding of the present magnetic field. The magnetic field outside a uniformly magnetized sphere is identical to that produced by a dipole, which can be thought of as a strong bar magnet, placed precisely at the center of a sphere, which otherwise is non-magnetic. No measurements can be made external to this sphere to distinguish between these two hypothetical sources. Worse yet, an infinity of other magnetized states can also produce the same result. We say the problem is non-unique. For that matter, we cannot even distinguish at one instant in time whether electric currents within our planet produce the magnetic field or whether the field comes from permanently magnetized material in Earth's interior. We have to find other ways (discussed in chapter 3) to do this. Questions concerning uniqueness did not enter into the discussions of the origin of the magnetic field until relatively recently. Certainly early scientists such as Gilbert did not consider such questions when they argued Earth itself was a large magnet. Their explanation at that time was simple and appeared to fit all the data. Even using modern scientific criteria, their solution would have been judged the best solution at that time. Of course it is also wrong, as scientists eventually learned. No consideration was given to the connection between electric currents and magnetism, because that connection was not discovered until the nineteenth century. Science is constantly evolving. It is questionable whether an ultimate scientific law is accessible to humans for any phenomenon. Probably every law should be considered an approximation of the rules governing our universe. Excluding a few missteps, scientific laws evolve by producing better approximations that we hope converge on some ultimate set of laws.

While not being able to determine the sources of Earth's magnetic field uniquely makes our task of learning about the origin of Earth's magnetic field more difficult, it makes our description of Earth's field easier. We can describe the vast majority of the present Earth's magnetic field as originating from a strong bar magnet located precisely at Earth's center. We can do this even knowing the actual sources of Earth's magnetic field are primarily produced from electric currents traveling in Earth's iron-rich core. (The reasons we know this will be given in chapter 3.)

If Earth's magnetic field were that depicted in figure 1.1, there would be no primary longitude singled out by the magnetic field, because the dipole at Earth's center is along the rotation axis. However, the dipole best explaining our present magnetic field is tilted at 10.3° with respect to the rotation axis. In 2005 the geomagnetic poles of this dipole were at 79.7°N, 71.8°W and 79.7°S, 108.2°E. If this were all there was to the magnetic field, the mariners would have been able to find a prime magnetic meridian separating the two hemispheres. One could designate the prime magnetic meridian as the longitude occupied by the north geomagnetic pole. But that was not to be.

Mariners encountered a variety of technical problems in navigating with a compass at sea. The first involved how to deal with the ship motions affecting the compass. A solution was to use a set of rings, each of which could pivot in a manner to keep the compass horizontal (a gimbaled system). In 1538 the Portuguese explorer João de Castro noticed a ship's gun causing a deviation in a compass needle. Mariners found by trial and error the distances iron objects had to be kept from compasses to allow for accurate navigation. The magnetic field of a gun, assumed here to have a dipole field, falls off rapidly as the cube of the distance. If the distance between the compass and the gun were doubled, the field strength would be decreased by a factor of 1/8. If the distance were tripled, the field from it would be reduced by 1/27 and so on. Lightning also occasionally reversed the polarity of some compasses, which then had to be remagnetized to restore their original polarity.

Deviations in the magnetic field associated with magnetization in crustal rocks can affect compasses, but not by as much as ancient mariners thought. Before Oersted found in 1820 that electric currents affected a magnetic needle, the only scientific explanation for magnetic phenomena involved permanent magnets. In the seventeenth and eighteenth centuries, this led to speculations of large magnetic mountains rising up from the depths of the ocean. Even today we recognize that some seamounts, submerged volcanoes, are so strongly magnetized that they affect magnetic compasses. The deviations caused by the magnetic crust to the main magnetic field, which originates in Earth's core, are called magnetic anomalies. The speculations of huge magnetic anomalies are common in history and folklore. In the second century AD, Claudius Ptolemy spoke of islands, now thought to be near Borneo, so strongly magnetic that ships with iron nails were forever caught in their grip if they passed too closely. A different version appeared in Arabian folklore. Magnetic mountains pulled the nails out of ships passing by, causing the ships to break up and sink. Still later so-called magnetic continents were argued to cause variation in declination, the deviation of the compass needle from true north. Even today, when people should know better, one occasionally hears speculations concerning unusually large magnetic fields alleged to cause the demise of modern ships in the neighborhood of the Bermuda Triangle.

The concept of magnetic mountains segued into the concept of multipoles: there were thought to be several places on Earth's surface where the magnetic field was vertical. The location closest to the north pole where the field is vertical is referred to as the magnetic north pole. In 2005 the magnetic north pole was at 83.2°N, 118.0°W, and the south magnetic pole was at 64.5°S, 137.8°E. Unlike the geomagnetic poles, the north and south magnetic poles are not 180° apart. Why is this, and why do the geomagnetic and magnetic poles not coincide? The answer is that Earth's magnetic field cannot be completely described as originating from a geocentric dipole field, a field produced by a strong magnet at Earth's center. The field remaining after we subtract out the dipole field from the total field is referred to as the nondipole field. That is, the total field is the sum of the dipole and nondipole fields.[12]

The nondipole field exhibits a more complicated spatial variation over the surface of Earth than the dipole field. In contrast to the dipole field, which is vertical only at two poles, the nondipole field can have several locations where it is vertical. A prime magnetic longitude dividing Earth into two hemispheres can't exist because a pure geocentric dipole field doesn't represent the total field.

Worse yet for the search for a magnetic Holy Grail was the discovery that the field changes with time. Henry Gellibrand (1597–1636), professor of astronomy at Gresham College, appears to have been the first to document that the magnetic field varies as a function of time. Gellibrand based this conclusion on measurements of magnetic declination recorded in London of 11.3°E in October 1580 (measured by William Borough), 6.0°E in June 1622 (measured by Edmund Gunter), and 4.1°E in June 1634 (measured by Gellibrand). The implications of this for navigation were not fully appreciated until later. It is not even clear that Gellibrand understood the implications of his findings or believed the magnetic field was changing with time. He could not rule out the possibility that the recorded differences in declination were due to inaccuracies in the measurements preceding his.

Reliable navigation required the accurate recording of magnetic field directions. Edmond Halley (1656–1742) became a driving force behind a magnetic survey at sea to produce the first magnetic charts. Beginning in 1698, he took a nearly two-year voyage on the sailing ship *Paramore*, which was lent to him by the British Admiralty for the purpose of making a magnetic survey of the Atlantic Ocean. Magnetic charts based on this survey, published in 1701, employed the first use of contour lines, which in this case connected points of equal declination, the deviation of magnetic north from true north. (The latter could be obtained from astronomical

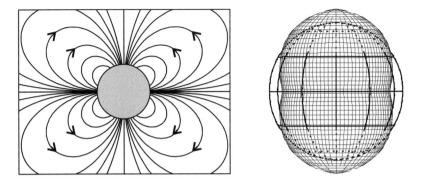

FIGURE 1.4 Axial quadrupole field. The field exits Earth at the equator and enters Earth at the poles as shown in the figure on the left. The figure on the right is a grid showing the inclination of this quadrupole field over the surface of Earth. It is positive (downward) at the poles and negative (upward) at the equator. This picture illustrates in three dimensions the symmetry of the field. (For geomagnetic specialists, where the grid in the figure is outside the circle, the inclination is downward and vice versa when it is inside the circle.) In particular, the field is outward at any location on the equator. (Mathematically, the axial quadrupole field is one of five quadrupole fields needed to completely characterize all second-degree spherical harmonics.) Figure provided by Phil McFadden.

data.) Throughout most of the rest of the eighteenth century, these contour lines were commonly referred to as Halleyan lines.

Halley was initially perplexed by the findings of Gellibrand. If Earth were permanently magnetized, how could the field vary with time? Eventually he thought he solved the problem: Earth had four magnetic poles. Two were on the surface of an outer sphere and two others were on a concentric inner sphere that rotated westward with respect to the outer one. This produced a "westward drift of the magnetic field," as it was to become known. This westward drift was supposedly facilitated by a fluid medium between the inner sphere and the hard shell of the outer sphere. Halley appears to have been very proud of this explanation, because he included a diagram showing the concentric spheres in a portrait of him made when he was eighty. Ironically, this explanation has been shown to be false, while many of his other contributions had more important impacts on science. As we shall explain later, the actual magnetic field is now thought to originate from numerous electric currents primarily concentrated between the depths of 2,900 km (about 1,800 miles) and 5,100 km (about 3,100 miles) in Earth's outer core.

Halley's four magnetic poles were described as a quadrupole field. However, this definition of the quadrupole field is no longer used. Figure 1.4 shows an example of a quadrupole field as now defined.[13] This field is vertical everywhere at the equator and at the two poles; it does not have four poles at Earth's surface. However, it can be non-uniquely represented as originating from two bar magnets (having a total of four poles) placed end to end at Earth's center and oriented along Earth's rotation axis. The quadrupole field is an example of a nondipole field. The modern-day terminology came from applications of a mathematical technique called Fourier analysis to the magnetic field data.

*   *   *

Most scientists and mathematicians have learned about Fourier analysis, a mathematical technique developed by the famous French mathematician Joseph Fourier (1768–1830). It is sometimes claimed that mathematicians make their most important contributions before the age of thirty. I don't know where this claim originated or whether there is any truth to it. Certainly, Fourier does not fit this description, perhaps because he was sidetracked from doing mathematics by Napoléon. Fourier was Napoléon's scientific adviser during France's invasion of Egypt in 1798. When the French fleet was destroyed in August of that year in the Battle of the Nile, Fourier had no way to return to France. Instead he served as an administrator during France's occupation of Egypt, where he helped in the development of various educational facilities. When he was able to return to France in 1799, he accepted another administrative position upon Napoléon's urging. He oversaw such activities as highway and irrigation development. He did not complete his most important mathematical memoir "The Propagation of Heat in Solid Bodies" until he was forty. Although this memoir apparently caused some controversy at the time, it is now highly regarded. It was in this memoir that Fourier used a mathematical procedure now referred to as Fourier analysis.

Fourier analysis is used for numerous reasons, including to provide quantitative estimates of periodicities in data. Consider a line that moves across a paper from left to right. This line can have as many wiggles as you like as long as it does not double back on itself. The mathematical technique developed by Fourier allows one to represent this line with a sum of the periodic trigonometric functions, sines and cosines. A single sine (or cosine) can be viewed as a wave in which the spacing between adjacent maxima, the wavelength, is constant. By applying Fourier series to our wiggly line, we can, for example, determine if any particular wave is dominant. The reason this might be important will be illustrated momentarily.

A different illustration of Fourier analysis can be obtained by considering music. A tuning fork tuned to middle C executes 261 vibrations per second: it has a frequency of 261 Hertz. The frequency of one octave above middle C is 522 Hertz, twice the value of middle C. Twice the number of compressions and rarefactions of air reach our ear from this note relative to the one an octave lower. This is the next higher-degree harmonic to middle C. Suppose we refer to middle C as a first-degree harmonic. The C one octave higher would then be a second-degree harmonic and so on. In principle, there are an infinite number of harmonics. Most sounds consist of a complex mixture of waves having different amplitudes and wavelengths. There are many notes and harmonics present. Fourier, or harmonic, analysis breaks down this complex mixture into a series of individual harmonics. By doing this, one can find which harmonics contribute the most.

To illustrate the usefulness of Fourier analysis, let's consider whether our climate has exhibited any periodic variation over the past few million years, a question we will return to in chapter 6. If our climate exhibits predictable cycles, we might be able to make future forecasts more reliable. For instance, we might better distinguish how much of the present climate warming is due to natural causes and how much is due to human input. To determine if cyclic behavior is present, we need to analyze data, such as proxies of climatic change recorded in sediments deposited on the ocean floor or in ice cores taken from the Greenland and Antarctica ice caps.[14] This area of research is referred to as paleoclimatology.

While most people know that greenhouse gases in our atmosphere influence our climate, other factors are also important, including variations in Earth's orbit. For example, the 23.5° tilt of Earth's rotation axis (the angle between Earth's rotation axis and a line perpendicular to the plane Earth orbits the Sun) determines our seasons. The radiation we receive from the Sun depends on the position of the North Pole; that is, whether our rotation axis is inclined in the direction of the Sun or away from it. When the North Pole points in the general direction of the Sun, we have summer in the Northern Hemisphere. When it points away from the Sun, we have winter.

The tilt of the rotation axis varies between 22° and 25° over a time interval of 41,000 years. Does this variation significantly affect the average annual temperatures and precipitation of our planet? In 1941 a Serbian scientist, Milutin Milankovitch (1879–1958), suggested it did. Furthermore, he suggested that other variations in Earth's orbit with periods of 23,000 and 100,000 years also significantly affected our climate.[15] These suggestions were not well received by the scientific community. The sparse data used by him to arrive at his conclusion were not of high quality. Calculations also indicated that the orbital changes were too small to have much effect on climate. Subsequently, the data have substantially improved and more sophisticated analyses have been applied. This allows scientists to determine if there are periodicities embedded in the paleoclimatic records recorded in ice cores and marine sediments. Applications of Fourier analysis to the data indicates that a dominant 100,000-year periodicity in temperature appears to have occurred during the past 900,000 years, while a 41,000-year periodicity dominated the climate during the preceding million years.[16]

Fourier analysis is an extremely valuable tool for scientists. But some artifacts can arise if it is not applied and interpreted correctly. For example, Fourier analysis can be used to show one periodicity is larger than another. But is it significantly larger? The use of the word "significant" in science usually means some statistical analysis has been applied to the data. If something is significant, it has a much higher probability of not having occurred by chance.[17] To illustrate what I mean, let's consider a simple problem for which Fourier analysis is not required. When a fair coin is flipped, it has an equal probability of landing heads or tails. Suppose after flipping a coin 100 times, we find 52 instances of heads and 48 of tails. Can we conclude that the coin is not a fair one because it seemed to have shown a bias toward heads? A statistical analysis, which can yield the probability that any particular number of heads will occur, indicates that the answer is no: more times than not, either heads or tails will dominate after 100 coin tosses.[18] Because similar, but more subtle, problems can occur in evaluating results from Fourier analysis, erroneous periodicities of various phenomena are sometimes published in peer-reviewed articles. We will encounter a few of these later in the book.

Often erroneous periodicities are difficult to get rid of once they have been published in a scientific journal. The media often interprets attempts to eliminate an erroneous result as reflecting controversy in science, when in reality it is sometimes nothing more than correcting an error. Unintentional errors in science are more common than a layperson might imagine. A statement often heard over coffee between scientists goes something like, "If a scientist hasn't been wrong, he has not published enough or he is far too conservative." Fortunately, if an error has serious consequences, it will eventually be discovered and corrected. This is the nature of science.

Once during a national radio talk show, the moderator proclaimed that cycles in the universe dominate our lives. She asked me to explain why periodicity was so prevalent in our universe. This threw me for a moment. It is important to know which questions to ask in science. I haven't the faintest idea how to answer the question "How cyclic is nature?" What parts of "nature" is she talking about? In contrast, Milankovitch's question, whether certain climatic variations were caused by variations in Earth's orbit, is a well-posed one. It can be answered by harmonic analyses of data.

Many people, including scientists, describe periodic behavior as an example of order. As some philosophers point out, it is difficult to define "order" without defining "disorder." Chaos is sometimes used to describe disorder. However, it is not always a simple matter to distinguish order from chaos. This is evidenced by deterministic chaos, a useful subject to know when we discuss magnetic field reversals.

Deterministic chaos is often known as the Butterfly effect: "Does the flap of a butterfly's wings in Brazil set off a tornado in Texas?"[19] The butterfly flapping its wings in Brazil is an example of an initial condition. Deterministic chaos is often described as involving phenomena very sensitive to initial conditions.[20] Cliff Mass, a University of Washington professor in atmospheric sciences, says sometimes weather conditions are difficult to forecast, while at other times they are easy; sometimes one is dealing with deterministic chaos and sometimes not. It is a complicated business. If we were trying to forecast the weather in Texas six months into the future and if deterministic chaos was involved, we might need to know at this moment all the factors that affect the weather across our entire planet. These factors would be our initial conditions. For simplicity, we refer to this as our initial state. Our weather forecasting abilities clearly depend on how precisely we need to know this initial state.

To illustrate dependency on the initial state, let's consider a far simpler example than weather. Consider any number between 0 and 10, say, 2. We will call this our initial state. We now establish a rule to describe how this state evolves with time. The rule I choose is to square the number. Thus, with time the state changes from 2 to 4, then to 16, followed by 256, and so on. Because of this rule, the evolution of this system is deterministic. We can precisely calculate how the state evolves. We have a recipe to determine the value of the state (number) after as many steps as we choose, providing we precisely know the initial state. For our choice, the original number (2) evolved rapidly to a larger number. But if we had started with ½, it would have decreased rapidly with each succeeding step: first to ¼, then to $\frac{1}{16}$ and so on. Only the number 1 (referred to as a fixed point) does not change with time. Consider the problem of determining the final state of some phenomenon obeying the above mathematical recipe when the initial value is infinitesimally close to 1. You carry out an experiment to determine its initial value. However, experimental errors occur in any experiment: you can't precisely determine whether the initial number is slightly less than 1, equal to 1, or slightly greater than 1. Because of this uncertainty, you cannot predict how the initial number will change with each step; for that matter, you can't even tell whether it will increase or decrease. If the initial number is infinitesimally smaller than 1, it evolves to a smaller number. In contrast, if it is infinitesimally larger than 1, it increases with time. Our rule is very sensitive to the initial condition (the number we initially chose). This is characteristic of deterministic chaos. The chaotic-like behavior occurs, even though I have provided you with a rule on how to calculate the value after every step. It is a deterministic process that leads to an unpredictable outcome.

It is interesting that astrologers talk about cycles in nature, while advocates of intelligent design, an offshoot of creationism, argue that nature is too complex for there not to be a God. It would be interesting to put these two groups of pseudoscientists in the same room to see how the battle develops. However, before scientists become too smug, we should recognize our views have altered substantially during the past two decades. For example, it has been common to use the clockwork cycles of the planets as an example of periodicity and order. The rotation of Earth about its axis and its orbit about the Sun were used as the standard timepieces for the day and year, until atomic clocks emerged on the scene. The Milankovitch cycles also use the regularity of minor variations in Earth's orbit. Periodicity seems to dominate planetary motions within our solar system. However, astronomers point out that all is not as it seems. They find that the orbits of many planets and moons are only approximately periodic. Over long intervals, more than 10 million years or so, they exhibit deterministic chaotic motion. This chaos arises from gravitational interactions between various astronomical bodies in our solar system. In principle the orbits of planets and moons can be precisely calculated far into the future by using the laws of mechanics, but in practice one cannot do this because the calculations are too sensitive to the initial conditions.

* * *

Gauss (1777–1855) was the first to apply harmonic analysis to describe Earth's magnetic field. Although his baptismal name was Johann Friedrich Carl Gauss, he signed his mathematical papers Carl Friedrich Gauss. Unlike Newton, Gauss was a humble scholar who appeared little concerned with receiving credit for his incredible number of accomplishments. His precocious nature was discovered shortly after birth when, before the age of three, he reportedly found his father's long payroll computations to be incorrect. Gauss's didactic memory allowed him to master several languages early in life. For two years after he entered the University of Göttingen, he remained undecided whether to pursue mathematics or linguistics (then called philology). By the time he entered the university, he had already developed the method of least squares, a valuable tool used widely in statistics.[21] The Gaussian law of normal distribution of errors, better known outside of scientific circles as the bell-shaped curve, was developed by Gauss and reflects his long-term interest in error analysis. He did not restrict his interests to statistics. Indeed, it is difficult to find subjects in mathematics Gauss was not interested in and to which he did not make major contributions.[22] Gauss also made important contributions to science, including mechanics and electromagnetism. One of the four Maxwell equations is often referred to as Gauss's law.

Gauss wanted to describe Earth's magnetic field, which is characterized by a direction (usually given by its declination and inclination) and a magnitude (intensity) at any location. He wanted to describe the field everywhere on the surface of Earth, even though there were only a finite number of measurements. He chose to use spherical harmonics, continuous functions defined on a sphere that have some of the same properties as sines and cosines. That is, he used a generalized type of Fourier analysis applied to a spherical (Riemannian) geometry to provide a quantitative way to extrapolate between measurements. We need not be concerned with the details, which are described in many mathematical books.[23] Gauss used three magnetic charts[24] to obtain 84 points. These points were fit with harmonics to describe Earth's magnetic field in 1838. In particular, he determined how much of the field is made up of the dipole field, the lowest-degree harmonic, and the nondipole field represented by the remaining

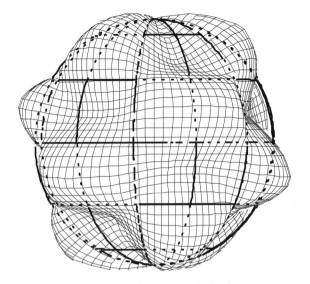

FIGURE 1.5 A more complex spherical harmonic field. This figure depicts the inclination of a more complex nondipole field than shown in figure 1.4. The degree of the harmonic is related to how many ups and downs the field makes with latitude. Not mentioned in the text is the fact that something called the order is related to how many ups and downs the field makes with longitude. In this case, the field is represented by a sixth-degree and third-order spherical harmonic. This terminology is not important for understanding the material given in this book. This figure is presented to illustrate that spherical harmonics can be used to represent complex magnetic fields. Figure provided by Phil McFadden.

harmonics. An example of a dipole field is shown in figure 1.1, and one of a nondipole field is shown in figure 1.4.[25] Higher-degree harmonics show more spatial variation than lower-degree ones, as illustrated by figure 1.5. This is also illustrated in figure 1.6, which shows the vertical component of the nondipole field at the turn of this century. The large number of spatial variations exhibited by the nondipole field illustrates that many higher-degree harmonics are needed to describe the present magnetic field of Earth. The more ups and downs Earth's magnetic field exhibits, the higher the degree of harmonics needed to describe them.

Gauss's mathematical description of Earth's magnetic field can be confusing when translated into words. It artificially places all the internal sources (but not external sources, to be defined in a moment) in Earth at its center. For example, Gauss's method would represent the field from a dipole offset some distance from Earth's center as coming from nondipole sources and a single dipole placed precisely at Earth's center. The larger the offset of a dipole is, the larger the magnitude and number of nondipole

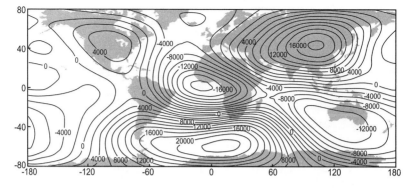

FIGURE 1.6 The vertical component of the nondipole field for the International Geomagnetic Reference Field for the year 2000. Contours of the vertical component of the field are labeled in units of nanoteslas. Figure drawn by Beth Tully.

terms needed to represent it. These results reflect the fact that Gauss's method cannot be used to locate uniquely the actual magnetic sources. Gauss found in 1838 that the dipole component of the magnetic field at Earth's surface strongly dominated the nondipole one. This has remained the case ever since.

Gauss's method also allows one to distinguish between internal and external magnetic field sources. Internal sources are defined as originating beneath Earth's surface, and external sources are ones originating above the surface. Lightning between clouds is an example of an external source. Lightning is an electric current and, as with all electric currents, there is an associated magnetic field.

In 1722 a London instrument maker, George Graham, noticed small changes occurring during the course of a day in a sensitive compass he had made. These daily changes are referred to as the diurnal variation, and they are now known to come from electric currents in Earth's upper atmosphere. Changes in electric currents in the upper atmosphere can sometimes have dramatic effects, such as those produced by magnetic storms. For example, in 1989 a powerful magnetic storm resulted in a power outage in Quebec, which affected several million people for nine hours. Although magnetic fields arising from magnetic storms can be relatively large for a week or so, the average annual external field is relatively small (chap. 4). As first found by Gauss, and confirmed by all subsequent analyses, less than 1 percent of Earth's magnetic field averaged over a year comes from external sources.

With time the data have increased and the methods of error analyses have improved. Nevertheless, Gauss's method is essentially the one used today. During the past century, direct measurements of Earth's magnetic field have been made continuously at permanent magnetic observatories. (At present, there are about 170 of these.) These are supplemented by various oceanographic measurements, portable land stations, and aircraft measurements. During the past few decades, several satellite missions have also been devoted to magnetic field measurements. In 2004 the first magnetic observatory on the seafloor was installed off the Washington coast. Measurements from these sources are used to obtain an average magnetic field for a given year, referred to as an International Geomagnetic Reference Field, IGRF. For example, the 2000 IGRF refers to the average field for the year 2000. Magnetic field values for an IGRF are given in the form of tables for the harmonics and contour maps of different magnetic elements, such as inclination and intensity. Following a decision made during the International Geophysical Year of 1967–1968, an IGRF has been produced every five years.

One can also describe how Earth's magnetic field changes with time by using harmonic analysis. To illustrate how this is done, consider a photograph of an ocean taken from a tower on a beach. It will show the position of waves at one instant in time. There will be waves with different wavelengths and different heights. One can take a photograph from the same location a short time later. The position of the waves will have changed. By knowing the time between the two photographs, one can estimate the velocity of the waves and how fast the wave shapes are changing. A quantitative estimate of this can be obtained by using harmonic analysis. A similar process works for Earth's magnetic field. For example, one can use two IGRFs from different times. By comparing the 1945 IGRF to the 2000 IGRF, for example, we can determine how the magnetic field has changed with time between 1945 and 2000.

The change over time of Earth's magnetic field is referred to as the geomagnetic secular variation. Not only do we have harmonic descriptions of Earth's magnetic field and its secular variation going back to the time of Gauss, but we have them going all the way back to the sixteenth century, well before Gauss carried out the first harmonic analysis of magnetic field data. About two decades ago, a student in England, Jeremy Bloxham, now a professor at Harvard, and his advisor, David Gubbins, analyzed data from manuals and logbooks of the ancient mariners to extend spherical harmonic models of Earth's magnetic field back to the early seventeenth

century. Subsequently, others have extended models back into the sixteenth century.

My abbreviated description of how some geomagnetists apply Gauss's
method to modern data to produce a magnetic field model, such as an
IGRF, makes the process seem rather simple. But it isn't, and politics
sometimes trumps science. For example, it costs money to run a magnetic
observatory, and one must be able to justify the costs. Some underdevel-
oped countries are proud to run an observatory and participate in an international science project, while others resent doing so. This means some
observatories have sufficient funds to produce first-rate results, while
other results are marginal. But if scientists declare a certain country's
data to be of low quality, this may result in the closing of their observatory. Then data from that region would no longer be available. This would
degrade future magnetic field models, because we know the best IGRFs
are produced when data come from widely distributed sites across our
planet. The International Association of Geomagnetism and Aeronomy
(IAGA) has committees to help resolve such problems. IAGA publishes
an "official" IGRF. However, individual scientists often disagree with the
political decisions on how the input data have been weighted in this IGRF.
Thus, they produce their own IGRFs, which compete with the official one.
Fortunately, the differences between these IGRF models are minor and
don't affect any conclusions given in this book.

It is curious that many scientists spend considerable time worrying
about details in data, sometimes out to several decimal points, but seem
unaware of how the data are used. This can be illustrated by an interaction I had with a scientist, whom I will refer to as Bob (not his real name)
during a meeting in Washington, D.C., in the early 1990s. This meeting
involved a new geomagnetic initiative. Scientists often participate in such
initiatives to learn new science and to gain an edge up on obtaining funds
for their research. During a coffee break I asked Bob, one of the world's
leading experts on magnetic field models such as IGRFs, what percentage
of the present Earth's magnetic field is nondipolar. He replied, "Around
10 percent." Because this seemed right, I went on to discussing other
things until I realized I might want to reference the source of this information in a future article. I asked Bob where he got the 10 percent nondipole
figure. He replied it was from a book I had coauthored with Mike Mc-
Elhinny in 1983. He was giving our number back to me! We had obtained
this percentage from other experts who made the estimate during the
1960s, a time when computing power was substantially lower. Often in

writing books, one doesn't recalculate all the numbers used. Instead, one often relies on results published in peer-reviewed scientific journals. In fact, a graduate student of mine, Reiner Heller, later found that the 10 percent estimate was far off the mark. Scientists in the 1960s didn't have the fast computers we have today, and presumably the original 10 percent estimate was no more than an educated guess at the time. Unfortunately, this 10 percent figure is sometimes still quoted in the twenty-first century. Even in scientific circles, it is not easy to expunge an erroneous conclusion if it has been cited enough times.

Heller found that the nondipole component constituted about 17 percent of the total intensity of the magnetic field at Earth's surface in 1900 and about 25 percent in 2000. Figure 1.6 shows the vertical component of the nondipole field at Earth's surface in the year 2000. While always dominant, the dipole field declined by 6 to 7 percent during this hundred-year interval.[26] I refer to these percentages at Earth's surface because the ratio of nondipole to dipole fields decreases with distance above the surface (see n. 25). The field becomes more dipolar the farther away we are from our planet. If we were very far away, Earth's magnetic field would essentially appear to be a pure dipole field. Similarly, if we could move downward toward the sources of the field in Earth's core, the nondipole fraction would increase in size.

Although both the dipole and nondipole fields change with time, the nondipole field appears to change quicker. In 1950 Sir Edward Bullard concluded that the nondipole field on average is moving westward at about 0.18° per year. Following Halley's lead, this is referred to as the westward drift of the nondipole field. Although this drift will be useful to us later when we discuss the origin of Earth's magnetic field, it also paints an overly simplistic picture. Some parts of the nondipole field even drift eastward. In addition, the nondipole field is largely absent throughout much of the Pacific region. The amplitude of the nondipole field decreases to near zero as it moves into the hemisphere containing the Pacific Ocean.[27]

Because the sources of Earth's magnetic field cannot be uniquely determined from measurements made at or above Earth's surface, it is difficult to determine the extent to which the nondipole and dipole field sources are intertwined. An electric current loop produces both dipole and nondipole magnetic fields.[28] That is, any individual current loop in Earth's core will make some contribution to both the dipole and nondipole fields. Yet the decline of the dipole field concomitant with the increase in the nondipole field during the past century suggests that some separation of dipole from nondipole field sources also exists. Our inability to identify the precise locations and contributions of sources has allowed some geomagnetists to model the field and its secular variation using dipoles offset from Earth's center. While sometimes useful for instructional purposes, such non-unique representations of the sources of the field are not favored by modern-day theorists, for reasons given in chapter 3.

The changes in Earth's magnetic field receive considerable media attention at times. Although both the north magnetic and geomagnetic poles are moving in nearly straight lines in a northwesterly direction, the north magnetic pole has moved the most during the past century. In 1900 it was at latitude 70.5°N, in 1950 it was at 74.6°N, in 2000 it was at 81.0°N, and at the time of this writing (2008) it was at 84.2°N. The south magnetic pole is not tied to the north one and is moving at a rate of around 5 km per year, while the rapidly moving north magnetic pole is heading northwest by about 50 km (31 miles) per year. A few geomagnetists have suggested that tourists who now go to Fairbanks, Alaska, to view the northern lights (aurora borealis) might be better off going to Siberia later in this century to see them (chap. 4). This prediction assumes the continuation of the present movement of the magnetic north pole. However, our knowledge of the present magnetic field and its secular variation is still too poor to conclude this with any reasonable degree of confidence. The magnetic pole could just as easily decide to go off in an entirely different direction over the next several years.

The change in the magnetic field receiving the most attention by media and geomagnetists alike is the demise of the dipole field, whose decrease has accelerated over the past 2,000 years, as we learn from the rock record discussed in the next chapter. This has led a few geomagnetists to predict we are heading toward a magnetic field reversal within the near future when the north and south geomagnetic poles will swap hemispheres. Could they be correct? What would be the consequences for humans if this were to happen? Before we can answer such questions, we need to examine the evidence for magnetic field reversals.

CHAPTER TWO

Magnetic Field Reversals

The geological history of Earth's magnetic field is written in rocks. As explained in this chapter, deciphering this record is not easy, and it has given birth to some of the major geological controversies of the past century. However, it was not clear at the beginning of my graduate school career that I would live long enough to see the resolution of these controversies.

*   *   *

I somehow managed to get about two-thirds of the piton into the rock before I had to grab it. I was hanging a thousand feet above Yosemite Valley's floor from a thin piece of metal. Fortunately, the rock is what climbers describe as "good rock"—in this case it was granite, an igneous (from the Latin *ignis*, or fire) rock that had formed over time from the cooling of magma far below Earth's surface. Some of the minerals (mostly quartz, mica, and feldspar) in this granite had grown large enough during the slow cooling that climbers could now use them to move up the shear walls of Yosemite Valley in California. We would sometimes pinch the larger mineral crystals between our first finger and thumb to get a hold. I had become tired doing this, and now I was in the precarious position of dangling from a quickly placed piton while fearing that my life might end

right then in the 1960s when I was barely in my early twenties. After a
moment's rest, I gathered my strength and finished the rock pitch.

A couple of years later, I was using a hand-carried rock drill to obtain samples from similar rocks in the Sierra Nevada mountains in order
to study Earth's paleomagnetic (ancient magnetic) field. Diamonds embedded in the drill bit allowed us to obtain rock cores approximately an
inch in diameter and several inches long. We drilled the rocks to increase
our chances of getting fresh rock that had not been altered by chemical
change.

I had already learned in my graduate school studies at the University
of California at Berkeley that direct measurements of Earth's magnetic
field, including those made by mariners with compasses, cover a very small
segment of geological time. If the 4.6-billion-year history of Earth were
compressed to a year, the length of time that we have direct measurements
of Earth's magnetic field would be roughly a second. Indirect measurements obtained from rocks were required to learn about the history of the
field throughout the vast majority of geological time. I was destined to drill
many rock cores and make years of measurements in a laboratory as part
of my PhD research at Berkeley. Less than a percent of the minerals in the
rocks I sampled were capable of becoming permanently magnetized. None
of the primary minerals used to identify granite are capable of doing this.[1]

Because permanent magnetization can change, scientists prefer to
use the term "remanent magnetization,"[2] of which there are many kinds.
We refer to the remanent magnetization acquired when a rock forms as
the primary magnetization. Usually it is so weak it doesn't even affect a
compass needle. However, on rare occasions, when magnetic minerals are
sufficiently abundant, rocks lying far below the sea's surface can produce
magnetic anomalies strong enough to affect mariners' compasses. Weakly
magnetized rocks, ones with magnetic fields millions of times smaller than
Earth's magnetic field, can now be routinely measured in scientific laboratories.

Although many different types of rocks are used in paleomagnetic
studies, paleomagnetists prefer rocks with a strong stable magnetization.
Some rocks, such as granite, are only occasionally used because their magnetization is often unstable: over time it can change its direction and intensity. The magnetization of some granites can even be changed by striking
them with a hammer. One rock commonly used by paleomagnetists is the
igneous rock, basalt. It is the most common lava flow on the surface of our
planet. Large floodplains of basalt occur in India, Siberia, South Africa,

and the northwestern United States, and basalt constitutes a major part of the oceanic crust. Because basalt cools quickly, its minerals have insufficient time to grow to a size to be visible to the naked eye; a microscope must be used.

Basalt and other rocks often contain small amounts of magnetite (Fe_3O_4), a mineral useful in paleomagnetic studies.[3] Magnetite, like all magnetic minerals, is not capable of carrying a remanent magnetization at high temperatures. When magnetite is heated, its magnetization decreases because of a phenomenon referred to as Brownian motion, a random movement produced by thermal fluctuations. Consider the heating of magnetite crystals in a rock. While the crystals do not move, the individual magnets in them do. When heated, these tiny magnets (atomic dipole moments) jump around like water droplets at the bottom of a hot pan. No remanent magnetization occurs when these dipole magnets are pointing in random directions: the magnetic fields they produce cancel each other out. The temperature above which a remanent magnetization is no longer possible is called the Curie temperature. It varies for different magnetic materials, which have different atomic structures. Magnetite has a Curie temperature of 580°C (1,076°F).[4] When magnetite is cooled to room temperature from temperatures higher than its Curie temperature, it acquires a remanent magnetization, referred to by specialists as a thermoremanent magnetization, which is the primary magnetization of igneous rocks.

Basalt magma becomes completely solid around 1,100°C (2,012°F), which is well above the Curie temperature of any known mineral. Therefore, basalt is not capable of acquiring a remanent magnetization until it cools far below the temperature that it becomes solid. Remanent magnetization is confined to Earth's thin crust because Earth's deeper interior is too hot for it to exist there, a result further discussed in the next chapter.

Studies of magnetization in rocks arguably began with one of the founders of modern chemistry, Robert Boyle (1627–1691). Although Boyle is best known for his pioneering studies of gases, he also delved into magnetism to show that bricks recorded the Earth's ambient magnetic field direction when they were fired and subsequently cooled. More than a century later, the Italian Macedonio Melloni (1798–1854) was the first to show that basalts also did this. However, the subject did not take off until Pierre David and Bernard Brunhes began their studies early in the twentieth century, as will be elaborated on later in this chapter.

The use of rocks to determine Earth's paleomagnetic field have provided a wealth of information. Nevertheless, paleomagnetic results have often been the center of controversy. Doubts about the fidelity of the rock magnetic record and interpretations of that record still sometimes occur, even though they are of a different form from those of a half a century ago when the derogatory term "paleomagician" was commonly used to describe a paleomagnetist at international scientific meetings.

Thousands of experiments now show that basalt accurately records Earth's paleomagnetic field. But in the developing stages of paleomagnetism, skeptics suggested that the rock record changed with time. A major controversy developed and peaked during the middle of the twentieth century after some English scientists, particularly Keith Runcorn and his students, claimed paleomagnetism could be used to prove that continents drifted over time. This was widely viewed as nonsense before evidence of plate tectonics, which requires continental drift, emerged. Most earth scientists, including nearly all those in North America, concluded that paleomagnetism must be wrong. When scientists don't like an idea, they search for something in the data or assumptions to criticize. Almost always some weakness emerges, especially when one is dealing with a subject at the frontiers of science. The skeptics quickly concluded that something must be wrong with the magnetic recording system. However, the data were too convincing to challenge the conclusion that primary magnetization often locked in the direction, and occasionally the intensity, of the magnetic field. In contrast, the claim that the primary magnetization retained the record of the paleomagnetic field for millions of years seemed absurd. Skeptics argued that the primary magnetization would decay over time and a new magnetization, a secondary magnetization, would be acquired after the rock formed. Sometimes this did happen. For example, a secondary magnetization was sometimes acquired when a rock underwent chemical alteration or when a rock was struck by lightning. The skeptics concluded that they could ignore paleomagnetism and continental drift in favor of the conventional wisdom of the day.

Scientists believe Earth has cooled and contracted over time. According to the scientific consensus in 1960, this contraction squeezed rocks until the crust buckled, forcing its weaker parts upward to form mountains. These mountains were then worn down by erosion. Tectonic processes, such as mountain building, were thought mostly to involve vertical movements of the crust. The horizontal movements advocated by drifters, those scientists who believed in continental drift, were incompatible with this.

A particularly outspoken and influential scientist critical of paleomagnetic arguments for continental drift was a Cambridge University professor, Sir Harold Jeffreys (1891–1989), who had already applied the laws of physics to prove it was "impossible" to move continental crust through oceanic crust. I first met Jeffreys when I was a student, and I learned the hard way what many scientists already knew: he was a genius, sure of his convictions, who intimidated those who did not agree with him. He was not alone when it came to doubting paleomagnetic results: most scientists had strong views on the subject, and most concluded that paleomagnetism was junk science.

By 1960 Sir Harold Jeffreys had been knighted for his many contributions to geophysics (the use of physics to study Earth), astrophysics, and statistics. Although at times shy, he was a broad scientist who published several editions of a book simply entitled *The Earth*. Throughout his career, he opposed the theory of continental drift and its successor, plate tectonics. He particularly disliked paleomagnetic evidence for drift. He often argued that the magnetic record was unreliable because the remanent magnetization could be altered by the simple process of a paleomagnetist using a geological hammer to break off a rock from an outcrop to bring it back to study in a scientific laboratory.[5]

Skepticism is an essential ingredient of science. The vast majority of scientists in the United States agreed with Jeffreys: they neither accepted continental drift nor that rocks could serve as reliable recorders of the paleomagnetic field. One promising paleomagnetist, John Graham, even left the field after he carried out experiments showing that pressure could change the magnetization in some rocks. This disillusioned scientist concluded that paleomagnetism must be wrong. The questions and concerns raised by Jeffreys and others in the middle of the twentieth century needed to be addressed. Fortunately, science is not done by consensus, and ultimately sufficient evidence accumulated for the minority point of view to become accepted—but not before many egos were bruised. The acceptance of continental drift had to wait until plate tectonics was accepted. Jeffreys held on to many of his views long after they were dismissed by mainstream science: he still believed continental drift and plate tectonics were nonsense when he died in 1989.

Although this is not the place to answer all the questions raised by skeptics, it is useful to provide an explanation to Jeffreys's question. The magnetic properties of minerals depend on many factors in addition to composition, such as the size and shape of the grains (see appendix). There are as many magnetic components in a rock as there are individual magnetic grains. Numerous experiments now indicate that some of these components can be reset by hammering (or by squeezing) the rock, but others cannot. The components reset by stress can be changed by applying a laboratory magnetic field about 200 times stronger than the magnetic field at Earth's surface. The magnetization affected by such a field is referred to as a soft magnetization. Paleomagnetists have used this information to develop laboratory techniques to remove the soft magnetization, leaving a hard magnetization that has a higher probability of retaining an accurate record of Earth's ancient magnetic field.

Although paleomagnetists now routinely carry out many experiments in the laboratory to determine the primary magnetization, on occasion errors still occur. Even today a few paleomagnetic results are contested because of doubts concerning the fidelity of the rock magnetic record. A problem is that not all hard magnetizations are primary. One way a hard secondary magnetization can be produced is through chemical alteration of a rock. For example, the red color seen in some rocks reflects the formation of magnetic iron oxide minerals by a process called weathering. Because weathering occurs after the rock formed, it is often impossible to determine the age of the magnetization. It doesn't help geologists to know the direction of the magnetic field in the past if they don't know when the field had that direction. It was because of weathering that I had used a drill to obtain fresh rocks in California during my PhD research: I wanted to avoid the surface of rocks that might have been altered through time by chemical weathering. The chemical alteration problem is complex because most rocks show evidence of some chemical alteration. Fortunately, not all chemical changes in a rock destroy the primary magnetization (see appendix).

Although paleomagnetists try to avoid using rocks in which secondary magnetic minerals have likely formed, on occasion mistakes still occur. Because of this, paleomagnetists routinely use a variety of tools to determine if the rock record is valid. One example comes from sedimentary rocks. When sediments settle in the ocean, magnetic grains align themselves along the magnetic field, much like a compass needle aligns itself to point to magnetic north. The grains accumulate on the seafloor, where they are compressed into a sedimentary rock. The primary magnetization acquired during settling and compaction involves a completely different mechanism from that of igneous rocks. (The mechanism by which igneous rocks acquire their magnetization is described in the appendix.) Paleomagnetists conclude that the rock record is more reliable when sedimentary and igneous rocks with similar geological ages from the same region provide consistent results. It would be surprising if the completely different mechanisms for acquiring a primary magnetization in igneous and sedimentary rocks yielded identical erroneous recordings of the paleomagnetic field.

* * *

Tens of thousands of rock samples have now been examined, and even if a small percentage of individual results are wrong, the rock record can provide us with a remarkably accurate picture of Earth's magnetic field in the past. Let's assume this is so and examine what the record tells us. Along the way we will explore additional things that scientists have learned about the magnetism in rocks, including circumstances when the rock record is far from ideal. We begin by discussing continental drift.

A commemorative stamp was issued in Greenland in 2006 honoring the Berlin-born scientist Alfred Wegener (1880–1930) for his explorations of Greenland. Wegener, a trained meteorologist, was an adventurer who once held the record for the longest flight (52 hours) in a hot-air balloon. His first trip to Greenland was in 1906. His second trip occurred in 1912 when he and the Danish explorer J. P. Koch took forty days to travel more than 1,100 km (682 miles) across Greenland's ice cap. Wegener died on his third trip at the age of fifty while returning from a lengthy trip to the middle of Greenland to supply a meteorology outpost. They found his body a year later, in 1931, wrapped in a sleeping bag. He was about halfway between the post and the coast.

With Wegener's death the world not only lost an explorer; it also lost a creative scientist. At a Frankfurt Geological Association meeting in 1912, Wegener speculated that continents drifted with respect to one another. Later he became a lieutenant in the German army and was shot in the arm during a German advance into Belgium in 1914. This, and a subsequent neck injury, provided him with enough convalescence time to develop his ideas on drift further. In 1915 he published a book, *The Origin of Continents and Oceans*, in which he proposed there once was no Atlantic Ocean: the North American and South American continents were joined with the European and African continents to form a supercontinent called Pangaea. Continental drift subsequently led to the separation of these continents.

There was considerable resistance to the acceptance of Wegener's theory of continental drift. One of his arguments for continental drift was that Africa and South America fit together like pieces in a jigsaw puzzle. He thought continents floated on a denser substrate and were forced through the oceanic crust by tidal and rotational forces. Jeffreys appeared correct: this mechanism was inadequate, and the data offered by Wegener supporting drift appeared weak. For example, opponents argued the fit between Africa and South America would not be impressive when sea levels were different enough to affect the shape of the continental margin. If the East Antarctic ice sheet were to melt today, virtually the entire state of Florida would be underwater. (We now know the climate was warmer when the Atlantic Ocean began to form: there were no continental ice sheets on Antarctica or anywhere else.) Why accept continental drift under such circumstances when the traditional explanation of most geological phenomena seemed satisfactory?

The existence of Pangaea was not accepted until after paleomagnetism became a respectable science, some time during the 1960s.[6] Scientists now believe that supercontinents formed and broke apart throughout much of Earth's past. Pangaea is the most recent and best documented of these, existing from 300 to 175 million years ago.

Doubts about paleomagnetism were not restricted to the fidelity of the magnetic record stored in rocks. Most scientists also did not accept a crucial assumption made by paleomagnetists that the magnetic field of Earth averaged over several thousand years can be represented as originating from a geocentric dipole aligned along Earth's rotation axis, as shown in figure 1.1. This has become known as the geocentric axial dipole field assumption. This was a controversial assumption and remains so even today for some older geological periods.[7] The assumption is built on geomagnetists' knowledge of the geomagnetic secular variation, the change of Earth's magnetic field with time. The magnetic field seems to change in such a way that non-rotationally symmetric components of the field might be averaged out over time. In particular, an average geocentric axial dipole field requires that the nondipole field and the tilt of the dipole field average to zero over time. This assumption was based on the possibility that Earth's rotation plays a crucial role in producing Earth's magnetic field, an assumption now accepted in modern dynamo theories, as discussed in the next chapter. The scientists making this assumption reasoned that the dipole at Earth's center should show an average preference for the rotation axis. They could not imagine any other axis in Earth's core that would be singled out for special consideration in the then-rudimentary dynamo theory. But such intuition requires testing. A few English scientists used the statistics of Ronald Fisher (1890–1962), arguably the most famous statistician of the twentieth century, to analyze the magnetic directions from rocks that formed within the past several thousand years. They concluded that the average magnetic field was a geocentric dipole one for the most recent 10,000 years. They assumed this was also the case for any similar time interval in Earth's past—a big extrapolation. Scientists, particularly Keith Runcorn and his students, used this assumption to find the positions of continents in the past.

When the geocentric axial dipole assumption is valid, the average magnetic field looks like that shown in figure 1.1. The north geomagnetic pole, called the paleomagnetic pole and obtained from averaging the primary magnetizations of different regions, coincides with the true north pole.[8] This can be used to find the ancient latitude, or paleolatitude, of the site where the average magnetization was obtained. When the geocentric axial dipole assumption is true, the direction of the average primary magnetization points toward the paleomagnetic pole, and the distance to the pole depends on the inclination of the magnetization. Consider a continent sitting on the equator 100 million years ago on which a series of lava flows erupted over, say, 1 million years. The average direction of magnetization of these flows would be horizontal because the continent sits on the equator. Now fast-forward to today and assume you can obtain the primary magnetization in a scientific laboratory. Because this magnetization shows no inclination, the 100-million-year-old paleomagnetic pole for this continent is at an arc distance of 90° in the direction the magnetization points.[9] If the continent had changed latitude during the past 100 million years (or if it had rotated about some axis through the continent), the paleomagnetic pole and present north pole would not coincide. The continent would have moved relative to the north pole during the past 100 million years. Paleolatitudes of continents can be obtained from the inclination of the average primary magnetization, and this can be used to test whether continental drift occurred. When different continents yield different paleomagnetic pole positions for the same time interval, one concludes that the continents have drifted (moved) with respect to one another.

In 1924 Wladimir Köppen (1846–1940) and Wegener proposed that India had drifted about 5,400 km (3,350 miles) northward (about a 49° latitude change). This suggestion was tested three decades later. One of Keith Runcorn's students, Ted Irving, measured many samples shipped to him from the Deccan Traps, a name given to a very large number of basalt lava flows that erupted in India around 65 million years ago. These lava flows erupted over a time span of a million years or so, which Irving assumed would be sufficient to average out the secular variation: the geocentric axial dipole field assumption should be valid. He found that the average magnetization direction of these rocks indicated that India had moved northward about 6,000 km (3,720 miles) during the past 65 million years. (The error in this estimate was given as plus or minus 1,000 km.)

Ted Irving, a student at Cambridge University in England, concluded that Wegener's theory of continental drift had been confirmed. But others did not. Irving failed his final PhD examination at Cambridge, where Sir Harold Jeffreys's intimidating intellect sometimes dominated earth science thinking. Irving was not able to convince the examination committee that his paleomagnetic evidence for drift was valid. Nevertheless, he was accepted on the faculty at the Australian National University, where he continued to do research in paleomagnetism. In 1964 he published the first book ever written on paleomagnetism.[10] After paleomagnetism and plate tectonics had become respectable, Cambridge University awarded Irving a doctorate of science degree.[11]

*   *   *

Not all paleomagnetists accepted the validity of continental drift during the 1950s and 1960s. One of the future stars of the field, Allan Cox (1926–1987), decided to investigate whether magnetic field reversals occurred because he doubted the paleomagnetic record could be used to test adequately the continental drift hypothesis. As we shall learn, Cox played a major role in establishing the magnetic field reversal chronology, which convinced many scientists of the reality of continental drift.

The hypothesis that Earth's magnetic field had reversed in the past preceded Wegener's hypothesis of continental drift. The French scientists Pierre David and Bernard Brunhes published a series of papers (sometimes together and sometimes separately) from 1901 to 1906 showing that several recent lava flows from France accurately recorded Earth's magnetic field when they cooled. Brunhes and David were surprised to find one flow (and its baked clay contact[12]) magnetized almost precisely opposite to the local direction of Earth's magnetic field. They could tell this flow was relatively old because the other flows, which erupted later, lay on top of it. This is a type of dating, called relative age dating, commonly used by geologists. It provides relative ages for rocks, but not absolute ages. It can be used to tell which rock is older than another, but it cannot be used to provide a quantitative estimate of how old the rocks are. David and Brunhes were puzzled by the magnetization recorded in the relatively old basalt because it differed by nearly 180° from the present field direction. Could it be possible that Earth's magnetic field had reversed polarity? That is, could the north and south geomagnetic poles have traded places? This speculation became known as the magnetic field reversal hypothesis.

Rocks that become magnetized in the present magnetic field have magnetic directions pointing northward, the same as a compass needle does. Such rocks are said to have normal polarity. Oppositely magnetized rocks, such as first found by Brunhes and David, exhibit reverse polarity. Because there was no historical evidence of magnetic compasses pointing in the opposite direction, if a magnetic field reversal occurred, it must have been before compasses were used for navigation.

Subsequently, Paul Mercanton in 1926 and Motonori Matuyama in 1929 carried out extensive studies of other lava flows and found strong evidence of reverse polarity in older rocks. In particular, Matuyama found that nearly half the numerous volcanic rocks from Korea and Japan had reverse polarity. Although he had no absolute dates for the flows, he speculated Earth's magnetic field had reversed near the beginning of the ice age. In spite of the seemingly mounting evidence for magnetic field reversals, the field reversal hypothesis was not widely accepted until more than a half century after David and Brunhes carried out their pioneering studies.[13] There was a distrust that rocks could serve as accurate magnetic recorders, and there was the possibility of self-reversal.

An unlikely role in this story was played by Patrick Blackett (1897–1974), who had received the Nobel Prize in physics for developing the cloud chamber and its use in discovering cosmic rays. He became interested in the origin of the magnetic fields in stars, and he thought, as did Albert Einstein, there might be a new physical law linking rotation to magnetism. After all, a "spinning" electron produces a dipole magnetic field. Perhaps the rotation of matter itself produced magnetic fields. This might explain the magnetic fields observed in planets and stars; all that had magnetic fields rotated quickly.[14] Through careful experimentation, Blackett disproved his own hypothesis. A sensitive magnetometer he developed failed to sense a magnetic field in a rotating 15.2 kg (33.4 lb.) mass of gold. He must have been a very influential person, as he borrowed the gold from the Bank of England. He published his negative findings in a classic paper in 1952. Although disappointing to Blackett, negative results are important to science. Moreover, the magnetometer he developed for his experiments became widely used during the 1950s and 1960s by paleomagnetists. Today this magnetometer has been replaced by others, such as cryogenic magnetometers, so sensitive that one must worry about contamination from substances such as weakly magnetized paint pigment.

During his study of magnetic fields in stars and planets, Blackett became aware of reversely magnetized rocks, and he asked his friend Louis Néel, who later received a Nobel Prize in 1970 for his work in magnetism,[15] could minerals ever self-reverse? Néel had already shown in 1949 an interest in paleomagnetism by explaining how certain magnetite grains, ones with diameters much smaller than the width of one of your hairs, could retain their magnetization for billions of years. Larger grains carried an unstable magnetization, one that was easily reset when the grains were put in a new magnetic field. With this work, Néel had established the theoretical foundations of rock magnetism. As was the case for many physicists, Néel contributed time to obstruct the Nazis in World War II. He developed a method to demagnetize ships that involved randomizing the magnetization in a ship's hull so that the ship's magnetization would be insufficient to set off mines. Less than six months after Néel developed his method, it had been used at several ports to demagnetize more than five hundred war ships. The demagnetizing of some military ships is still done today by maritime powers.

Néel answered Blackett's question in papers published in the early 1950s: there are mechanisms by which a self-reversal of magnetization can occur.[16] About the same time Néel was showing how a rock could acquire a magnetization opposite to Earth's magnetic field, two Japanese scientists found one that did.

Seiya Uyeda was a student of Taseki Nagata (1913–1991) at the University of Tokyo during the 1950s. Nagata had a long interest in the magnetic properties of rocks, and he wrote the first textbook on the subject. It was only natural that Uyeda's PhD research would involve magnetic measurements on rocks. Uyeda heated and cooled rocks to determine how reliably they recorded Earth's magnetic field. He was surprised to find that one igneous rock from the Haruna Volcano acquired its magnetization precisely opposite to the external field. Both Uyeda and Nagata were perplexed by this finding. They were unaware of the theory being developed by Néel. (Uyeda and others eventually showed that the details of the self-reversing mechanism for this rock differed from those proposed by Néel.[17]) No longer could one doubt the reality of self-reversal—it could be produced in a laboratory. Upon this discovery, some scientists, including Uyeda,[18] suggested there was no need to retain Earth's magnetic field reversal hypothesis: why invoke two mechanisms when only one is needed?

The near simultaneity of Néel's theoretical work with the finding of a self-reversing rock heated up the debate on field reversals. Although Blackett became a powerful advocate for self-reversal, he also described how one could distinguish between the self-reversal and field reversal hypotheses at a conference held in Israel in 1954 (see n. 13). He postulated that if self-reversal is rare and field reversal is the dominant mechanism for producing reverse-polarity rocks, rocks of normal polarity should exhibit different ages from those of reverse polarity. In contrast, if self-reversal is the dominant mechanism, there should be normal polarity rocks with the same age as those with reverse polarity. By describing this, Blackett again exemplified the qualities of a good scientist. He described a method to distinguish between competing mechanisms—a distinction successfully tested during the 1960s that showed he was wrong in his strong advocacy of the primacy of self-reversal.

A pivotal role in this story was played by Jean Verhoogen (1912–1993), a University of Brussels scientist who had written an article in 1939 for *National Geographic Magazine* entitled "We Keep House on an Active Volcano." Verhoogen had spent several months camping on the flanks of Nyamlagira Volcano in Zaire (then the Belgian Congo), the first of many field seasons he spent studying volcanic activity in the rift zone of East Africa. He was the first to measure accurately the gas composition of an erupting volcano and among the first to use mathematical thermodynamics to describe the evolution of volcanic rocks. His height of nearly six and a half feet, sharp intellect, and deep voice intimidated most professional scientists. This intimidation could occur in any of the four languages Verhoogen spoke fluently. After accepting a faculty position at the University of California at Berkeley in 1947, he changed the spelling of his first name to John so that Americans would pronounce it correctly.[19] When Verhoogen became agitated, the smoke puffs emerging from his ubiquitous pipe increased in frequency, an observation not lost on his students. Although students also found Verhoogen intimidating, they appreciated his excellent and stimulating courses. Unlike the vast majority of faculty in North America, he presented both sides of the continental drift and magnetic field reversal controversies. Verhoogen also advocated the same method as did Blackett in his lectures to distinguish between self-reversal and field reversal. While Verhoogen favored self-reversal, two of his PhD students, Dick Doell (1923–2008) and Allan Cox, were among the handful of scientists destined to prove him wrong.

Two groups of scientists emerged in a race to establish the reality and ages of magnetic field reversals. Some members of these groups overlapped at Berkeley. Both groups ultimately developed the same plan: obtain magnetic field directions and ages from basalts to distinguish between field reversal and self-reversal, as first suggested by Blackett and echoed by Verhoogen. If self-reversal was rare, one could determine when magnetic field reversals occurred in the past. One could develop a magnetic field reversal chronology.

The first group consisted of Doell, Cox, and another graduate of Berkeley, Brent Dalrymple, who specialized in radiometric dating. Doell, who had survived the famous World War II Battle of the Bulge and "incurable" cancer was a genius in designing instruments, including some magnetic instruments still used today. He received his PhD in 1955, the same year Cox began his graduate studies at Berkeley. Cox, who had served in the Merchant Marine and later in the U.S. Army, carried out the bulk of the theoretical analyses, including most of the statistical arguments used by this group. He was a scholar whose scientific presentations were described by many as "magical." It surprised no one when he later became president of the American Geophysical Union and a dean at Stanford University. Doell and Cox agreed to cooperate on the study of magnetic field reversals in 1958 while Doell was teaching summer school at Berkeley. Shortly thereafter both joined the U.S. Geological Survey (USGS) at Menlo Park, California, where they subsequently recruited Dalrymple to their team. Dalrymple used the potassium-argon dating technique (to be described shortly) to date geologically young lava flows—ones that had erupted within the past few millions of years.

The leader of the second group was Ian McDougall, who, after finishing his PhD at the Australian National University (ANU), spent a year at Berkeley learning the same age-dating technique Dalrymple was learning for his PhD dissertation research. Shortly after he arrived in Berkeley, McDougall met Dalrymple, and they quickly became good friends and remained so after McDougall returned to ANU. The second member of the Australian group was Don Tarling, who left England to carry out geomagnetic secular variation research under Ted Irving at ANU. Although Irving had already made important contributions to continental drift theory, he wanted to better evaluate the geocentric axial dipole assumption. At first neither McDougall nor Tarling was particular interested in establishing a magnetic field reversal chronology, in spite of urging from Irving to do so.

The USGS group published the first magnetic field reversal chronology that included "absolute" (radiometric) age dates in the journal *Nature* in 1963. They reported that the field has had normal polarity, as it does today,

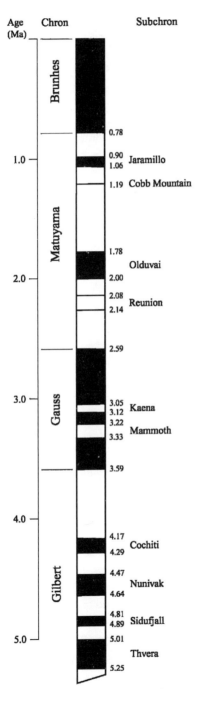

Age (Ma)	Chron	Subchron
	Brunhes	
1.0		0.78
		0.90 Jaramillo
		1.06
		1.19 Cobb Mountain
	Matuyama	1.78
2.0		Olduvai
		2.00
		2.08
		Reunion
		2.14
		2.59
3.0	Gauss	3.05 Kaena
		3.12
		3.22 Mammoth
		3.33
		3.59
4.0		
		4.17
		Cochiti
		4.29
	Gilbert	4.47
		Nunivak
		4.64
		4.81
		4.89 Sidufjall
5.0		5.01
		Thvera
		5.25

FIGURE 2.1 Reversal chronology. Normal polarity is in black, and reverse polarity is in white. The numbers refer to millions of years before present (Ma). The names Brunhes, Matuyama, and so on represent intervals during which one polarity dominates (now referred to as chrons), while the names Jaramillo, Cobb Mountain, and so on represent reversal events (subchrons). From NASA images.

for the past million years. Between 2 and 1 million years ago, it had reverse polarity, and before that it again exhibited normal polarity. Allan Cox once told me he believed this chronology was correct, because it illustrated that Earth's magnetic field had periodically reversed polarity. Cox was aware of data, discussed in chapter 4, demonstrating that the Sun's dipolar magnetic field reversed polarity approximately every eleven years. Although this eleven-year quasi-period was much shorter than Earth's apparent 1-million-year period, many scientists, including Cox, thought there might be a common mechanism explaining the magnetic fields of the Sun and Earth. However, the apparent periodicity of reversals of Earth's magnetic field was destined to disappear.

After McDougall realized they had more data than the USGS group had on this subject, he and Tarling responded to the USGS paper by submitting a manuscript to *Nature*, which was also published in 1963. Although the ANU group agreed with the USGS one that magnetic field reversals had occurred, they found no evidence of periodic reversals in the data. The race to establish a reliable reversal chronology was on. In retrospect, the differences in the two chronologies, and in many others to follow (both from these groups and from others who entered the fray), reflected the low number of flows sampled. As more and more data emerged, it became clear that Earth's magnetic field randomly reversed polarity, a conclusion later arrived at by Cox (and fortified by others), who carried out statistical analyses of the data.

The strength of the case for reversals was boosted after Neil Opdyke and his colleagues measured the primary magnetization in many marine sediments: the primary magnetization in these sediments was consistent with the reversal chronology established from lava flows. The validity of magnetic field reversals was widely accepted in the earth science community within three years after the first publications on the reversal chronology appeared in *Nature*. By then, the detailed measurements on rocks by the USGS and Australian groups, and others, had demonstrated that the phenomenon of self-reversal was real but rare. Figure 2.1 shows a recent estimate of the times the magnetic field reversed polarity during the past few million years.

*   *   *

Radiometric dating played a crucial role in demonstrating the reality of magnetic field reversals. Radioactivity is a form of transmutation that occurs when an atom spontaneously breaks down into two or more components. Henri Becquerel (1852–1908) received the Nobel Prize in physics in 1903 for his discovery of radioactivity in uranium in 1896. A fundamental understanding of the decay process was developed shortly thereafter by the team of Ernest Rutherford (1871–1937), who received the Nobel Prize in chemistry in 1908, and Frederick Soddy (1877–1956), who received the Nobel Prize in chemistry in 1921.

An example of a radioactive isotope is potassium-40, where the number following potassium refers to the sum of the number of neutrons (21) and protons (19) in its nucleus.[20] Potassium-40 decays (breaks down) to calcium-40 and argon-40 with a half-life of around 1.25 billion years. This means 1,000 atoms of potassium-40 will be reduced to 500 after 1.25 billion years, then to 250 atoms after another 1.25 billion years passes, and so on. As this decay occurs, the amounts of calcium and argon increase.

Knowing the half-life of an element is crucial in allowing geologists to obtain absolute age estimates of rocks. Usually this is determined from well-designed laboratory experiments measuring the rate of decay of radioactive atoms. Today there are many radioactive isotopes used for dating, including those of uranium, thorium, potassium, rubidium, carbon, and beryllium. However, the potassium-argon technique used by Dalrymple and McDougall was particularly well suited for magnetic field reversal chronology studies because it could be applied to basalt lavas that formed within the past few millions of years. These lavas were found to have lost their argon when they erupted at Earth's surface. After cooling, the argon accumulated in the rock over time. This could be used to calculate the initial amount of potassium-40 the rock had. By knowing the initial and final amounts of potassium-40 and knowing the half-life of potassium-40, geologists could obtain an estimate of the age of the rock.[21]

To understand how radiometric dating works, consider a bucket of water with a small hole in its bottom. The time it takes for half of the water to leak out of the hole is its half-life. If we start with a full bucket and after 15 minutes the bucket is half full, the half-life is 15 minutes. Suppose the bucket has a red color and the water leaking out of it is collected in a blue bucket. The amount of water initially in the red bucket is equal to the total amount of water in both buckets at any instance in time. If at one instant the red bucket is 1/8 full and the blue bucket is 3/8 full, how much time expired since the water began leaking out of the red bucket?[22]

Radiometric dating works in a similar fashion. For the potassium-argon method, one needs to know the amount of potassium-40 present when the lava first cooled (the amount of water initially present in the red bucket) and the final amount of potassium-40 (the amount of water in the blue bucket). By knowing the half-life, one can calculate the former from the amount of argon-40 in the rock, which is obtained by measuring how much is released when a rock sample is heated in a scientific laboratory. Because the fossil magnetization in some rocks is also locked in upon cooling, this dating technique is ideal to determine the times when Earth's magnetic field switched polarity.

Radioactivity is also used to determine the age of Earth and the solar system. Brent Dalrymple, of the USGS team involved in establishing a magnetic field reversal chronology, became upset by the way evolution was treated in primary and secondary schools in the United States and elsewhere. He served as an expert witness supporting evolution and radiometric dating at various trials, and he worked tirelessly for many years to provide scientific education to the general public, especially those members serving on school boards who make decisions on the teaching of evolution in public schools. He pointed out that age-dating methods, including the potassium-argon method, evolved with time. Although scientists made mistakes and corrected them in some of their age dating, this should be regarded as a healthy side of science rather than being used to show that science should not be trusted. Along the way Dalrymple discovered that many scientists also didn't know how Earth's age was determined. He responded by writing a book entitled *The Age of the Earth*.[23]

The age of the oldest known rock (a gneiss from Canada) is 4 billion years. However, minerals (zircons) within some rocks from Canada and Australia yield ages close to 4.4 billion years. These observations indicate that Earth's crust was altered after it formed. Most of the older rocks were destroyed, but a few of their mineral constituents survived and were incorporated in rocks that formed later. Dalrymple reviewed a variety of isotope techniques to conclude that Earth is 4.54 billion years old. The oldest known Moon rock (an anorthosite from the lunar highlands) has an age of around 4.4 to 4.5 billion years, while the oldest known meteorites have ages of around 4.57 billion years. Scientists believe the older age of meteorites relative to Earth and the Moon is significant: it took 30 million years or so for the terrestrial planets, including Earth, to form from the original dust cloud containing meteoritic material. Although I expect some revisions of these age estimates will occur in the future as more data becomes available and techniques improve, earth scientists would be shocked to learn Earth was, say, only 4 billion years old. In this book, I commonly refer to Earth's age as 4.5 or 4.6 billion years old, which spans the range that earth scientists are confident includes Earth's actual age.

Although the oldest reliable age of a rock containing a primary magnetization is around 3.5 billion years, most earth scientists suspect Earth had a magnetic field as soon as its core formed, probably within 50 million years after Earth itself came into existence. Some opponents of evolution have argued that Earth's magnetic field is inconsistent with these estimates. The intensity of Earth's magnetic field has been declining since 1838, when Gauss used spherical harmonics to estimate the character of Earth's magnetic field. These opponents extrapolate this decrease back in time to conclude that Earth's magnetic field would have been unbelievably large 10,000 to 20,000 years ago. Some even claim that this puts an upper limit on the age of Earth: it couldn't be older than 10,000 to 20,000 years.[24]

Such arguments are incompatible with magnetic field observations. Although Earth's magnetic field would have long ceased to exist if only decay were occurring, scientists recognize that the same mechanism that produced the field in the past is operating today. According to a well-developed mathematical theory, two mechanisms can produce temporal changes in Earth's magnetic field: (1) The field decays with time, and (2) the field is altered by the flow of electrically conducting fluid in Earth's core. (The origin of Earth's magnetic field will be discussed much more in the next chapter.) If the second mechanism can be ignored, Earth's dipole magnetic field decays (dies away) with a half-life of around 12,000 years. However, when pure decay occurs, the mathematics requires that both the nondipole field and the dipole field simultaneously die away. But this does not happen: the intensity of the dipole field has been documented to have decreased by 6 to 7 percent during the previous century, while the nondipole field intensity increased (chap. 1) Therefore, some process, which scientists attribute to a dynamo, must be acting to maintain Earth's magnetic field.

A similar conclusion can be reached by considering paleomagnetic estimates of Earth's magnetic field in the past. Although it is more difficult to obtain the intensity of Earth's magnetic field than the direction of the field from rocks (appendix), paleomagnetists find convincing evidence that Earth's magnetic field intensity exhibited many maxima and minima during the past few hundred thousands of years. For example, while the intensity reached a maximum 2,000 years ago, it had a minimum value, about 40 percent lower than today's intensity, about 6,000 years ago. If only decay were occurring, there could never be any increase in Earth's magnetic field intensity, contrary to observations that the field increased between 6,000 and 2,000 years ago. Earth's magnetic field cannot be a remnant of a field acquired when the core was born. Instead, some active process, now identified as a dynamo (chap. 3), must continually be producing magnetic field energy.

Unfortunately, the evolution of science has been so great during the past few centuries that we have entered an age of specialization, which makes it more difficult to respond quickly to pseudoscience arguments. The magnetic field theory mentioned above is not known by the vast majority of scientists, including earth scientists and physicists. In the first part of the nineteenth century, some scientists claimed to know all of science, math, engineering, and philosophy. While such claims were probably ill advised, no rational person would even begin to consider making a similar claim in the twenty-first century. An example of the degree of specialization present today is evident in the physics department at the University of Washington. Faculty sit in one of twelve different research groups. It is rare when a faculty member in one of these groups has a good understanding of the research developed in another group. For example, a nuclear physicist will have considerable difficulty explaining new developments in condensed matter (solid and liquid) physics and vice versa. In math it is even worse, as I learned when I served on a university committee evaluating recommendations for tenure. It is common for the vast majority of members on the math faculty to not understand the mathematical theory published by a candidate being considered for tenure. Instead they often rely on a handful of authorities at other universities to evaluate the details, and sometimes even the impact, of the candidate's work. It is for such reasons that scientists insist research be published in peer-reviewed scientific journals. This does not mean we always agree with all the material published, but it does filter out considerable nonsense that otherwise would be published. This also illustrates the danger of accepting "science" published on the Internet, since it is usually not peer-reviewed.[25]

* * *

Let us return to the subject of magnetic field reversals, which have also been used to date the age of Earth's crust.

I occasionally ask students, "What is the evidence for magnetic field reversals?" Usually they reply that striped magnetic anomalies in the oceans demonstrate the existence of magnetic field reversals. In reality, a magnetic field reversal chronology was first established and then used to explain the origin of these "stripes." But what are these "stripes"?

The longest mountain range on our planet, around 80,000 km (49,600 miles) in length, is the mid-ocean ridge. Although this ridge does not always lie in the middle of the ocean, it does in the Atlantic, where it was first identified and named during the 1957–1958 International Geophysical Year by Marie Tharp and Bruce Heezen, two oceanographers at Lamont Geological Observatory (now Lamont-Doherty Earth Observatory). Along with topography, oceanographers occasionally measured magnetic anomalies, deviations from the main magnetic field originating in Earth's deep interior, caused by magnetized rock on the seafloor. Because these deviations were typically only a few percent or less, they towed their measuring instrument, a magnetometer, behind the ship to reduce the impact the magnetic field of the ship had on their measurements. The navy was particularly interested in their results, presumably because the data would make it easier to locate submarines.[26] Oceanographers, such as Arthur Raff and Ron Mason, found a series of magnetic anomalies paralleling, and centering on, the mid-ocean ridge. Typically the largest anomaly, about 3 percent of the main field, occurred above the mid-ocean ridge, and the strength of the anomalies decreased in size with distance from the ridge. We can liken these anomalies to a magnetic bar code on a product in a store: the bar code has magnetized stripes, which can be read when passed under a magnetometer. Ship captains soon recognized these stripes, which had been numbered by oceanographers, could be used for locating where they were. They were heard to say things like, "We are now over anomaly 6 and moving away from the ridge."

Arthur Raff and Ron Mason puzzled over the origin of these anomalies. Although occasionally offset by faults, the anomalies extended thousands of kilometers along the mid-ocean ridge system. They also exhibit a remarkable symmetry centered on the ridge. This symmetry can be illustrated by a hypothetical example. Suppose an oceanographer starts at a point above the ridge and travels perpendicular to it for a hundred kilometers to find herself above anomaly number 7. She could have traveled just as easily in the opposite direction from the ridge and found anomaly 7, which would have appeared as a mirror image to the one on the opposite side. The anomalies are symmetric about a line along the center of the ridge.

These long and symmetric magnetic anomalies were a puzzle to everyone who looked at them. When I was a student in the early 1960s, Mason visited Berkeley and gave a talk on the subject. Afterward, a few of us students accompanied several senior scientists to a lounge. Mason had placed a large map showing these anomalies face-up on a table, and everyone puzzled over their origin. No one could provide a good guess, let alone a reasonable explanation, for their origin. I still wonder why this was so, because this meeting occurred while the U.S. Geological Survey team was developing their first magnetic reversal chronology, which was to be the key to unlocking this mystery. Moreover, the meeting was attended by Cox, Doell, and Verhoogen (and other talented scientists, such as Perry Byerly, a world-famous seismologist), who played important roles in establishing the chronology. But the vision to unlock the door to this mystery didn't come from anyone in that room.

Harry Hess suggested in scientific talks that the upper part of Earth spread away from mid-ocean ridges. Later Robert Dietz published a paper on the subject in 1961 in which he coined the phrase "seafloor spreading." However, the geological evidence used to support seafloor spreading was viewed as equivocal, and the speculation received little support. In 1962 Drummond Matthews returned to Cambridge University after completing a magnetic survey over a mid-ocean ridge (the Carlsberg Ridge) in the Indian Ocean. He found a new PhD student, Fred Vine, waiting for him. Matthews had only received his PhD the year before, and he grappled with what problem to suggest to Vine. After some contemplation, he recommended that Vine explain the origin of the magnetic anomalies Matthews had found in the Indian Ocean. This is not an easy problem because the crust could be magnetized in different ways and still produce the same magnetic field anomalies measured at the surface of the sea. That is, the problem did not have a unique mathematical solution. After working on the problem for a while, Vine suggested that the crust might have recorded Earth's magnetic field reversals in a manner similar to a magnetic tape recorder. (Vine was aware of work on magnetic field reversals and Dietz's paper on seafloor spreading.) Vine's suggestion received only a lukewarm reception from Matthews. It combined two wild speculations: magnetic field reversal, then unproven, and seafloor spreading, a speculation dismissed by many at Cambridge University. Matthews's view changed after Cox, Doell, and Dalrymple published their reversal chronology paper in 1963 in *Nature*. Within only a month or two of this publication, Vine and Matthews had produced a paper for *Nature*, which hypothesized that the striped magnetic anomalies occur because of seafloor spreading. They suggested that magma erupted at ridges and cooled to record Earth's magnetic field. As spreading perpendicular to the ridge occurred, this lava

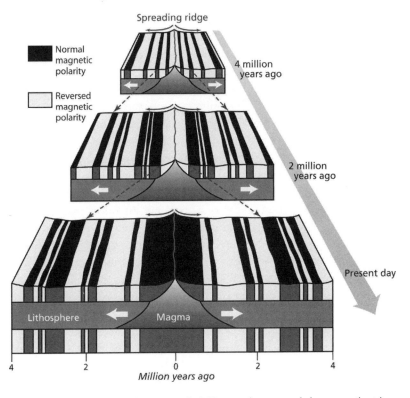

FIGURE 2.2 This figure shows how magnetic field reversals are recorded as magnetic stripes during the process of seafloor spreading. Figure provided by Maurice Tivey.

moved sideways to allow new magma to come in and record the magnetic field. A rock magnetized in a normal polarity field reinforces Earth's present magnetic field. The total magnetic field, the sum of the field originating in the core with that coming from the magnetized crust, is larger over a normally magnetized part of the crust; the resulting anomaly is then said to be positive. Since a reversely magnetized rock points in the opposite direction from Earth's main magnetic field, it produces a negative anomaly. Their interpretation is shown in figure 2.2. This speculation was called the Vine-Matthews hypothesis for many years, before becoming the Vine-Matthews-Morley hypothesis.

Lawrence Morley, of the Geological Survey of Canada, independently came up with the same idea as Vine, which he summarized in a letter submitted to *Nature* in 1963 (before the USGS reversal chronology paper

appeared). It was rejected, as was a second submission to a more specialized journal, the *Journal of Geophysical Research* (*JGR*). Morley told the
historian William Glen that an anonymous referee had concluded: "Such
speculation makes interesting talk at cocktail parties, but it is not the sort
of thing that should be published under serious scientific aegis."[27] Dick
Doell also read the manuscript and told me it was not as well written or
as well documented as the Vine-Matthews paper published in 1963. After Morley's manuscript was rejected by *JGR*, he teamed up with Andre
Larochelle to publish an article on the subject in 1964 in an edited book.[28]

While I was a student in 1965, I attended a large scientific meeting in
Washington, D.C. I chanced upon Larochelle sitting on a couch in a hotel lobby looking rather depressed. I asked him what was wrong, and he
replied that no one appreciated his work. I took the opportunity to make
some flattering comments about a more specialized paper of his; I was not
aware then of his work with Morley. As my comments seemed to make him
more depressed, I excused myself and walked toward the opposite end of
the large lobby, where Vine, surrounded by scientists and scientific reporters, expounded on the Vine-Matthews hypothesis, one of the most influential hypotheses in earth science of the previous century. I still don't know
what role Larochelle played in the paper he published with Morley. I suppose I never will, because Larochelle died before the hypothesis became
commonly known as the Vine-Matthews-Morley hypothesis. However, I
suspect Larochelle would not have been happy with the new name.

The seafloor spreading hypothesis required more time for acceptance than did magnetic field reversals, particularly at Columbia University's Lamont Geological Observatory (now Lamont-Doherty Earth
Observatory), where Maurice Ewing, a strong anti-drifter, was director.
However, Ewing was a strong believer in letting data tell the story. Although Lamont stored more than half of the world's magnetic anomaly
measurements and three-quarters of all marine (deep-sea) sediment cores
in 1965, little had been used in scientific publications. This bothered Ewing, who hired Neil Opdyke to carry out magnetic measurements on the
marine sediment cores. Opdyke, a geologist, had been a student of Keith
Runcorn and later spent a year with Ted Irving in Australia. During this
time he had become convinced that continental drift occurred. He found
little support for drift from the other faculty at Lamont, including Jim
Heirtzler, who led the marine magnetic anomaly group. Heirtzler and various colleagues had published papers in 1965 offering alternative explanations to the Vine-Matthews(-Morley) seafloor spreading hypothesis.

Following the publication of these papers, Heirtzler decided to analyze a very long profile of magnetic anomalies extending from the mid-ocean ridge in the south Atlantic to the coast of South America. He found that the magnetic stripes varied in size, including some very narrow ones. He thought this might be inconsistent with the seafloor spreading hypothesis and showed the data to Opdyke, who immediately opined that Heirtzler had just proved seafloor spreading. Heirtzler's data indicates that spreading of new seafloor west of the mid-Atlantic ridge occurred at 4.5 centimeters per year (slightly less than 2 inches), a rate comparable to how fast your fingernails grow. It is a slow rate, but sufficient to form the entire Atlantic Ocean seafloor in less than 120 million years.

Opdyke also saw a pattern of magnetic reversals in Heirtzler's data remarkably similar to one he was finding in marine sedimentary cores. In particular, Opdyke (and others[29]) had found support for a suggestion that relatively long intervals of constant polarity were punctuated with relatively short intervals of opposite polarity. None of these occurred since the last reversal happened around 780,000 years ago.[30] Between 2.58 and 0.78 million years ago, the field had reverse polarity, except for some relatively short intervals of normal polarity called reversal events. Some reversal events appear to be as short as 10,000 years or so. Events are not restricted to having normal polarity: reverse polarity events also occur during otherwise long intervals of normal polarity. Figure 2.1 shows some of these events.[31]

As an aside, Lamont scientists, such as Lynn Sykes, later used earthquake location data to demonstrate that plate tectonics was occurring. Earth consists of seven major plates and many minor plates that move relative to one another. These plates sometimes collide and give rise to mountain chains. For example, the Himalayas began forming about 50 million years ago when India moved northward to collide with Eurasia. However, most plates, which are formed at mid-ocean ridges when they separate and new magma erupts, are "destroyed" at subduction zones. As plates move away from spreading centers, they cool, contract, and become so dense that they break under their own weight and sink back into Earth. Subduction zones are about 100 km (62 miles) wide and descend at an average angle of 45° into Earth. These zones are essentially the only places in the world where earthquakes occur at depths greater than 75 km (47 miles): earthquakes occur as deep as 690 km (428 miles) in some subduction zones. The plate tectonics model requires seafloor spreading and continental drift. (It is discussed more in chapter 6.) But let's return to seafloor spreading, which provides a wealth of information on magnetic field reversals.

*   *   *

In just one year (1968), Jim Heirtzler and his colleagues had extended the magnetic reversal chronology from a few million years to 80 million years. Subsequently, others have extended it much further back in time. The oldest marine magnetic anomalies are around 160 million years BP (before present), a reflection that older oceanic crust has been subducted. Marine (and other) sediments and igneous rocks on continents have been used with the marine magnetic anomaly data to construct a reliable chronology for the past 150 million years. Only continental samples are used to establish older chronologies. However, large gaps often exist in the chronologies prior to 150 million years BP because reliable primary magnetizations are not always available from rocks. Sometimes rocks of the correct age are not available, and even when they are, secondary magnetization often has erased the primary magnetization. The reliability of the magnetic reversal chronology generally deteriorates further back in time.

Often the geomagnetic field reversal chronology for the most recent 150 million years is described by specialists as an "excellent first-order chronology." This translates to "There will be continual small changes in the chronology as more data are forthcoming." Nevertheless, the dating of magnetic reversals during the past 150 million years is now considered accurate enough for magnetic reversals themselves to be used to date rocks. For example, when a geologist locates the uppermost reversal in marine sediments, she would say the age of the sediments at the reversal point is around 780,000 years. This reflects the fact that the age of the last reversal has already been well determined from numerous measurements made elsewhere. In contrast, the reversal chronology for the time preceding 150 million years or so is a "second-order chronology." A translation of the loose language used here is "Don't be surprised if there are large changes in the chronology as more data accumulate." Prior to around 350 million years ago, the chronology is spotty and very poorly established back to the beginning of the Phanerozoic,[32] around 544 million years ago. The magnetic chronology prior to 544 million years ago is essentially unreliable.

Analyses of the magnetic reversal chronology reveals that hundreds of recorded reversals have occurred randomly over time.[33] The frequency at which reversals occur also changes over time. I am often asked what is the average interval of constant polarity. This is not a simple question to answer, in spite of common statements such as those given in some recent reference books, such as *The [Earth's] Magnetic Field Changes Direction [Reverses] Every 250,000 Years*.[34] I suspect *National Geographic* was trying to say in this quote that the *average* (mean) for the *past 25 million years* is around 250,000 years. The average value includes reversal events only 10,000 to 20,000 years in duration along with longer polarity intervals, such as the one we are presently in, which has already lasted 780,000 years. Earth's magnetic field does not reverse periodically every 250,000 years, as was erroneous implied by *National Geographic*. (For example, although the average of 640,000, 10,000, and 100,000 years is 250,000 years, one should not conclude a change occurred every 250,000 years.) Moreover, the average duration of a polarity interval between 50 to 25 million years BP is close to 600,000 years. This average is significantly longer than during the most recent 25 million years. Such observations made me curious about how Earth was regulating the rate at which reversals occur. This curiosity led to joint research with scientists in Australia.

During the mid-1970s, I was granted a sabbatical leave to work with Mike McElhinny at the Australian National University. It helped to initiate research with McElhinny to characterize Earth's paleomagnetic field and its secular variation using mathematical methods pioneered by Carl Friedrich Gauss in the 1830s (chap. 1). This ushered in a joint research program with McElhinny, which lasted throughout the rest of the twentieth century.

Mike McElhinny had written the second book ever published on paleomagnetism and tectonics in 1973. After a few years of doing research together, McElhinny suggested we write a book together on Earth's magnetic field, which we completed in 1983. One day while we were working on this book, he informed me that he was trying to convince Phil McFadden to come to ANU. At the time, I did not know who McFadden was.

Phil McFadden carried out paleomagnetic studies under David Jones in Rhodesia, Africa, during the 1970s to obtain a PhD in physics. Upon completion of his PhD, McFadden joined the faculty at the University of Rhodesia, where he might have remained had it not been for McElhinny and a civil war. Conditions for carrying out research were excellent in Australia and marginal in Zimbabwe, the name given to Rhodesia after the civil war ended in 1980. McElhinny described McFadden as a brilliant scientist who was also a genius in statistics. My first reaction was one of alarm: would there be anything left for me to do with someone like

McFadden around? About a year later, McElhinny arranged a three-week trip into the outback of Australia to help one of his students carry out research to determine how electrically conducting the Australian crust was. This student had four faculty as field assistants: McElhinny, Jones, McFadden, and me. McFadden and I stayed up late by ourselves every night of that trip arguing about everything imaginable. By the end of the three weeks, we had become good friends and had initiated joint research, which has continued up to the present.

We began by carrying out statistical analyses of the reversal chronology. We found that the probability of a reversal occurring has been increasing, on average, ever since 83 million years ago. Indeed, the probability of a magnetic field reversal is higher for the most recent 25 million years than any other preceding 25 million years for the past 150 million years, the interval of time for which a reliable reversal chronology has been established.

There were no reversals between 118 and 83 million years BP. This long polarity interval, recorded in marine magnetic anomalies and in terrestrial rocks, is referred to as a magnetic superchron of normal polarity. Whenever one finds a primary magnetization in a rock in this 35-million-year interval, it has a normal polarity.[35] The observation that Earth's magnetic field had normal polarity during this superchron does not reflect a bias toward normal polarity. Our analysis indicates polarity superchrons occur when the probability for reversal essentially vanishes. Whatever polarity state Earth's magnetic field happens to be in at such a time is the state it remains in until the probability for reversal increases. A 50-million-year-long superchron, which began 312 million years BP, has been found with reverse polarity.[36]

McFadden and I found these superchrons, first documented by others,[37] to be much longer than expected. Of the hundreds of well-documented polarity intervals during the past 350 million years, fewer than ten last more than 2 million years. Excluding the two superchrons, all polarity intervals are less than 4 million years. Based on detailed statistical analyses, we concluded that different dynamic processes acted in Earth's deep interior (to be discussed in the next chapter), separating times of superchrons from times when reversals were relatively common (on a geological time scale). The magnetic field of Earth can be divided into two mega-states. The first mega-state is one in which reversals are relatively common—as they have been for the past 80 million years. The second mega-state is one in which reversals did not occur.[38]

*   *   *

More than a hundred magnetic field reversals have occurred just since the dinosaurs went extinct 65 million years ago. Surely, one might think, scientists can explain what happens during a reversal transition. But that's not the case. The reason for this can be traced to the nature of Earth's magnetic field.

There are different ways to estimate how long it takes for a magnetic field reversal to occur. One method, used by Neil Opdyke and his colleagues at Lamont-Doherty Geological Observatory, required estimating the rate at which sediments accumulated on the ocean floor. A core was driven vertically into the bottom of the ocean to retrieve sediments. The primary magnetization in the top of the core was found to be normally magnetized. Farther down the core, the sediment was reversely magnetized. Between the normally and reversely magnetized sections of the core lay a transition zone recording the reversal of Earth's magnetic field. By determining the sedimentation rate (usually from relative age-dating techniques using fossils), Opdyke estimated the duration of the magnetic field reversal. About four decades ago, he (and others) applied this technique to many cores to conclude that Earth's magnetic field reversed polarity over a time span of a few thousand years. He also concluded that the intensity of Earth's magnetic field markedly declined during a reversal.

Subsequently, the methods of estimating the duration of reversals have increased, and many igneous rocks and sediment samples have been used to estimate how long it takes for Earth's magnetic field to reverse. The vast majority of estimates for the duration of the last reversal transition, which occurred around 780,000 years ago, fall between 1,000 and 10,000 years. We have produced no better estimates in the twenty-first century than scientists did decades ago! If anything, our range of estimates for this reversal transition has increased over time, contrary to the expectation it would decrease as more data accumulated. Let's see why.

When faced with a problem, theorists often construct a model. It makes sense to start with a simple model and, only after it is understood, to progress to more complicated models. The simplest model for Earth's magnetic field is to assume that it is a pure axial dipole magnetic field. That is, we assume that Earth's magnetic field comes from a bar magnet at Earth's center aligned along Earth's rotational axis. We (initially) neglect the nondipole field, even though we know it will be there in any real

situation. Even with this assumption, we are faced with a range of possibilities for describing a magnetic field reversal. One extreme model is that
a reversal occurs by rotating the central dipole through 180°. This means
the geomagnetic north pole moves along a longitude from the geographic
north pole to the geographic south pole. In this model, the intensity of
the dipole is kept constant during the rotation. In the second extreme
model for reversal, the dipole remains along the rotation axis but changes
in intensity: its size (intensity) decreases to zero and then increases with
the dipole facing in the opposite direction. The two extreme models are
shown in figure 2.3.

The transitional magnetic fields of the two models can easily be distinguished. There is no change in intensity in the first model, and the geomagnetic pole moves along a precise path from north to south. In contrast,
there is no path along which the geomagnetic pole moves in the second
model, and the intensity vanishes in the middle of the transition. Intermediate models between these extremes can also be distinguished. If the
dipole both rotates while its intensity decreases, it will move along a lon-
gitudinal path like the first model, but it will have a lower intermediate intensity like the second model. This can be determined, at least in principle,
by measuring the primary magnetization's direction and intensity.

So far, so good. Let's now construct a more complicated model by adding in a simple nondipole field. We know Earth's present magnetic field
consists of a dipole field and many different higher harmonic fields, which
we group together as a nondipole field. However, we still want a simple
model, and so we use one harmonic to represent the entire nondipole
field. We model the total magnetic field as the sum of the dipole and quadrupole fields shown in figure 2.4. We assume the quadrupole field remains
constant; the reversal occurs by a change in intensity of the dipole field in
this model. This is the same as the second extreme model previously considered, except now we have added in a quadrupole field, which we assume

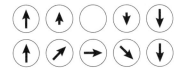

FIGURE 2.3 Extreme models for reversals. Circles in the bottom row show a reversal of a pure
dipole (indicated by arrows) field by rotation. Circles in the top row show a reversal when a
pure dipole field reduces its intensity to zero and then builds up to point in the opposite direction. Figure drawn by Beth Tully.

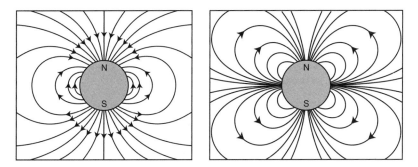

FIGURE 2.4 Reversals do not occur simultaneously over Earth. An axial dipole field is shown on the left, and an axial quadrupole field is shown on the right. When the polar intensities of these two fields are equal, the field vanishes at the south geomagnetic pole but not the north one. The N and S indicate the north and south geographic poles. Figure drawn by Beth Tully.

is smaller than the dipole one at the onset. How is this reversal process viewed by observers at the north and south poles, who are in telephone contact? (After all, this is a theoretical model.) Consider a point in time when the quadrupole and dipole fields have identical intensities at the north and south poles. Naturally the observers see only one field, which is the sum of the dipole and quadrupole fields. The direction of the quadrupole field is inward at the south pole (the arrow is into Earth), while the dipole field is outward. Because the arrows (representing the fields) are in the opposite directions but equal in size, the total magnetic field vanishes at the south pole when the intensities of the dipole and quadrupole fields become equal. The observer at the south pole calls the observer at the north pole to inform him they are right in the middle of a reversal transition. The observer at the north pole replies with, "You are nuts. There is a strong downward-pointing magnetic field here, as has been the case for a long time." The observer at the north pole observes the total field obtained by adding the arrow from the dipole field, which is directed into Earth, to that of the quadrupole field, which is also directed into Earth. The two fields add together at the north pole, but they cancel each other at the south pole.

The actual nondipole field is expected to be far more complicated than the one used in our simple model. Although we consider Earth's magnetic field to have reversed polarity when the dipole field reverses, the precise time when a reversal occurs cannot be determined from present data. As might be expected from the simple modeling given above, scientists at different localities will obtain different estimates for the duration of a magnetic field reversal when more realistic nondipole fields are incorporated into models for magnetic field reversal. We cannot even obtain a good average estimate for the duration of a reversal unless we have detailed knowledge of the character of the nondipole field throughout a reversal transition. We are very far from being able to do this because there are far too few detailed records of Earth's magnetic field during any particular reversal and because of age-dating uncertainties.

The maximum intensity drop during a transition also depends on the location at Earth's surface at which the field is being measured. The paleomagnetic record shows drops of intensity of Earth's magnetic field by as much as 90 percent. However, such low values of intensity probably occur when the nondipole and dipole fields are pointing in opposite directions. Most paleomagnetists believe the dipole field drops in intensity to about one-quarter of its usual value during a magnetic field reversal. I know of no paleomagnetist who thinks the intensity of the dipole field ever vanishes during a reversal.

There is more we can learn from our simple models. It is possible for an observer at one location to think a reversal has occurred while an observer at another location does not. Consider the simple model when the two observers at the poles disagreed about the timing of a reversal. At a point in time when the dipole has normal polarity but is smaller than the quadrupole field, the observer at the south pole concludes that Earth's magnetic field has reversed polarity, while the observer at the north pole disagrees. Suppose the dipole subsequently returns to its original direction and intensity. That is, the reversal process has been disrupted. The dipole has not been allowed to change sign to point in the opposite orientation in this model. After the dipole returns to its initial state, the observer at the north pole concludes that no reversal ever occurred. In contrast, the observer at the south pole claims a reversal event occurred. They both publish papers on their observations, which being well-cited, lead to increased funding and tenure at major universities.

Following a long interval of debate, scientists eventually recognized that a "local magnetic field reversal" can be recorded at some localities on Earth's surface, but not at others. When this happens, we say there has been a "magnetic field excursion." Some investigators hypothesize that an excursion is an aborted attempt of Earth's magnetic field to reverse polarity. In any case, several excursions appear to have occurred since the last reversal. This accounts for older claims published in scientific journals of one or more reversal events within the past 100,000 years. Now we think there are none. In contrast, the number of excursions since the last reversal at 780,000 years BP is estimated by different scientists to be between three and twelve.

Hundreds of magnetic field reversals have occurred randomly in Earth's past. Because some nondipole field is probably present during a reversal, it is difficult to estimate the duration of a reversal and the intensity change during a reversal. The likely presence of a nondipole field also makes it difficult to determine if different reversal transitions exhibit different characteristics. Although many of us intuitively expect some differences will emerge, such as differences in the rate that reversals occur, we did not expect the very rapid change in directions recorded in a 15.5-million-year-old flow in eastern Oregon state. Based on observations there, some paleomagnetists suggest that a change of 60° in the direction of Earth's magnetic field occurred in less than a year.

Approximately 15.5 million years ago, a thick sequence of lava flows erupted in northeastern Oregon at Steens Mountain. Three scientists from California—Ed Mankinen, Sherman Grommé, and Rob Coe—teamed up with a French scientist, Michel Prévot, to conduct some of the most detailed measurements ever made on a sequence of flows that erupted during a magnetic field reversal. Their work, initially published in 1985, was expanded on over the next decade by Coe and Prévot. They estimated the rate of change of Earth's magnetic field by calculating how much time it took various lava flows to cool and lock in the magnetization. Consider tracks made in wet cement by fast- and slow-walking beetles. A faster-crawling beetle will make a longer track before the cement hardens. Similarly, a faster-changing magnetic field will exhibit more total directional change in a lava flow than a slower-changing magnetic field, because the primary magnetization is locked in during the cooling of the lava. One lava flow, which cooled in less than a year, exhibited very rapid changes in direction. Coe and Prévot concluded that the magnetic field had changed in this flow at the astonishing rate of 6° per day and continued for several days. They only concluded that Earth's magnetic field could change very rapidly during a magnetic field reversal; they did not conclude that an entire reversal could occur in less than a year. However, others did. A few scientists and many laypersons extrapolated the author's results to claim that Earth's magnetic field could reverse polarity in a time interval as short as one month.[39]

The Coe and Prévot's results were not well received by many scientists.

Paul Roberts, one of the world's leading dynamo theorists, was quoted in a 1995 *Science News* article as saying, "To a theoretician like myself, these results are almost inconceivable." Roberts was referring to rates at which reversals occur in dynamo theories. (Dynamo theory is discussed in the next chapter.) Other scientists claimed that experiments, inferences of the mantle's composition (also discussed in the next chapter), and electromagnetic screening theory make it impossible to observe such rapid changes in Earth's magnetic field in the core, even if they did occur. Some of you may notice that your radio signal is affected when you drive across a bridge held up by metallic spans. The metallic suspension is a good electrical conductor and filters out some of the high-frequency radio waves. Similarly, the (semiconducting) mantle is thought to filter out all magnetic field variations with periods of less than several months. The shortest documented period of any magnetic field variation of core origin is around a year. Analyses of magnetic field records for the past century indicate that all shorter periods have been filtered out.

Not surprisingly, several alternative interpretations were made for the rapid changes recorded at Steens Mountain. One possibility was that the lava flow recorded an external magnetic field, one that, say, originated from electric currents in the upper atmosphere. However, the strength of the field needed and its duration made this explanation doubtful. The explanation that appears to be accepted by the majority of paleomagnetists was proposed by Mike Fuller (now at the University of Hawaii), who wrote a commentary on the subject for the journal *Nature* in 1989. Fuller pioneered paleomagnetic work on the character of magnetic field reversal transitions, and he is one of the planet's foremost rock magnetists (scientists who study the origin and properties of magnetic minerals and rocks). He suggested that the magnetization was not a primary one in the lavas showing rapid changes at Steens Mountain: the magnetization might be a secondary magnetization acquired over a long time interval after the lavas had cooled. Fuller appreciated that the magnetization used by paleomagnetists typically resides in grains so small they cannot be seen with an optical microscope. Subtle changes of the grain's size, shape, or chemistry are difficult, and sometimes even impossible, to detect using present technology. Yet such changes can significantly alter the magnetic direction and intensity over time (see appendix).

Although Coe and Prévot had previously found no evidence that secondary magnetization had affected their results, they returned to Steens Mountain with a PhD student of Prévot's, Pierre Camps, to collect two more lava flows in the reversal transition region. After analyses of the new data from these flows, they concluded in a 1999 *Journal of Geophysical Research* paper that the validity of the rapid magnetic field change hypothesis now seemed less likely. However, they also were unable to find evidence of a secondary magnetization artifact.

How should we interpret the results obtained from Steens Mountain? Scientists often say extraordinary claims must be supported by extraordinary evidence. It would be an extraordinary finding if rapid changes of Earth's internal magnetic field occurred in less than a year. It would overturn data and analyses of Earth's electrical conductivity structure, and it would require substantial changes in our view of how Earth's magnetic field is produced (chap. 3). I believe more plausible explanations, such as an undetected secondary magnetization, are responsible for the changes in magnetic directions recorded at Steens Mountain. It appears likely that no magnetic field reversal completely reversed in a time span substantially less than a thousand years.

*   *   *

Every ten years or so since magnetic field reversals were first documented, some scientist suggests that the present magnetic field is in the initial state of reversal. Combining such a prediction with claims that magnetic field reversals can occur very rapidly ("as evidenced by the paleomagnetic data from Steens Mountain") provides considerable ammunition for disaster scenarios.

Let's consider the evidence a few twenty-first-century scientists have given that we are in the initial stage of a magnetic field reversal. The dipole field is estimated to have decreased by 30 percent during the past 2,000 years. This decrease has recently accelerated, as evidenced by a 6 to 7 percent decrease in the intensity of the dipole field during the past century. It has been 780,000 years since the last reversal, while the mean time between reversals for the past 25 million years is around 250,000 years. We are overdue for a reversal, or so the proponents claim. These sound like compelling arguments, don't they? Why then do scientists like me urge caution?

Suppose you flipped a fair coin many times, and it just happened to come up heads several times in a row. What is the probability the next time you flip the coin it will come up heads? If it is a fair coin, as assumed, the answer is 50 percent. The coin does not know its past history. Although the statistical analysis used is different when treating reversals from that used in treating the multiple flips of a coin, Earth's magnetic field essentially also does not know its past history. (I say "essentially" because Earth's magnetic field probably does retain a memory for a few thousand years, as is discussed in the next chapter.) The present magnetic field does not know what polarity it had 50,000 years ago, let alone what it did 780,000 years ago.[40] The conclusion that we are overdue for a reversal does not imply that the probability for a reversal has increased.

Earth's magnetic field has exhibited many variations in intensity. The recent intensity has been either average or above average for the past million years (appendix). Many times in the past, it has been significantly lower and no reversals followed. For example, it was about 40 percent lower than today's field 6,000 years ago. Numerous minima in intensity have occurred since the last reversal. A few of these minima are probably associated with magnetic field excursions, but the vast majority are not. Not one of these minima has been followed by a complete reversal of polarity.

Because the field is almost always increasing or decreasing, the trick is to figure out some way to tell when any particular decrease might lead to a reversal. Paleomagnetists are investigating various properties of the nondipole field and possible relationships of it to the dipole field to determine if they can reasonably predict when the next reversal will come. Unfortunately, we only poorly know the character of the nondipole field preceding or during a magnetic field reversal. Thus, no one has been able to produce a viable method of predicting reversals. For that matter, it is not clear we can ever predict a reversal much before it happens. It may be a little like trying to forecast weather months in advance when conditions of deterministic chaos apply (chap. 1).[41] Nevertheless, changes in Earth's magnetic field, including reversals, will almost certainly occur in the future.

*   *   *

There are several possible consequences of a magnetic field reversal, as I will discuss from time to time throughout the remainder of this book. Perhaps the most dramatic of these is the possibility that some plants and animals might go extinct, a subject I was introduced to when I joined the University of Washington faculty.

One of the first scientists I met after joining the faculty in the late 1960s was the paleontologist Hsin-Yi (Jim) Ling. He was interested in small planktonic animals (radiolarians) and plants (silicoflagellates) that first appeared in the fossil record when the Precambrian ended about 544 million years ago to initiate the Phanerozoic eon. This is a time when hard shells, in this case made of silica (SiO_2), first became widely apparent in the fossil record. While plankton (Greek for "drifter" or "wanderer") are essentially carried about by oceanic currents, some species swim upward by more than a hundred meters in a day. When the plankton studied by Ling die, their siliceous hard parts sink to the ocean floor and are incorporated in sediment deposits. Oceanographers, such as Ling, take cores from the ocean floor to reconstruct the life and death of different species over time. If a species lived during a more recent time interval than a second one, its remains would be found at a shallower depth in the sediments than those of the second one. Over many years paleontologists have constructed a record of ancient life in the oceans.

Applying similar techniques, Neil Opdyke, Dennis Kent, and others also used sediment cores to help in the development of a magnetic field reversal chronology. Such work, still ongoing today, requires a close relationship between paleontologists and paleomagnetists. Ling had hoped I would become interested in deep-sea sediment cores and work with him: he would do the fossil studies, and I would measure the magnetic polarity.[42] Ling had already published a paper in which he argued a greater number of his siliceous fossils disappeared immediately following a reversal of Earth's magnetic field than would be expected from chance. He thought there was a connection: a polarity reversal somehow triggered extinctions of some species. He was not the first well-qualified scientist to report such findings and to speculate that reversals led to the elimination of some animal species from our planet.

Several mechanisms linking reversals to species extinctions have emerged over time. One of the first to appear involves the collapse of the Van Allen belts (chap. 4), which was speculated to increase radiation and extinctions. It soon became apparent that even if the magnetic field vanished during a reversal (it doesn't, as we have seen), the radiation in these belts would not reach the animals studied by Ling. Our atmosphere and the first centimeter of ocean water would absorb the radiation. Other mechanisms soon surfaced, including some rather clever ones. As discussed further in chapter 4, when the intensity of Earth's magnetic field decreases, more carbon-14 is produced in our upper atmosphere. Perhaps, after being incorporated into living organisms, the radioactive decay of carbon-14 produced genetic mutations leading to extinctions. However, because the amount of carbon-14, which has a half-life of 5,730 years, in our atmosphere is tiny (about one part in a trillion), this speculation seems unlikely.

Another possible mechanism involves the disruption of animal activities, such as migration, by changes in Earth's magnetic field, as discussed further in chapter 5. Small changes in animal behavior might trigger other changes leading to extinctions.

A key question here is how fast do environmental changes, such as changes in Earth's magnetic field, occur? Often the *rate* of change is more important than change itself. This is illustrated by an unfortunate remark made by Michael Griffin, the head of NASA, in 2007 (for which he subsequently apologized). While commenting on global warming, he said something to the effect that one should not expect any particular climate (such as the present one) to be the "best one." In other words, perhaps a warmer climate would be better. Although I doubt we could arrive at a consensus as to what constitutes a "best climate," Griffin had a point: it seems improbable that the present climate is the "best one." Nevertheless, Griffin was unwise in his comments, because he did not appreciate that it is the rate at which our climate is changing that is the problem. We would be far less concerned if the rate of global warming was much slower than observed (chap. 6). For example, if sea levels were to rise by a meter during this century, valuable coastline property in Florida would be lost. (A much larger effect would occur in the Maldives, a small island country in the Indian Ocean with the lowest high point in the world.) However, if it took a thousand years for the same sea level rise, few people would be concerned. We can more easily adjust to a rise of 10 cm (about 4 inches) than a meter rise during this century. It is the rate of global warming, rather than global warming itself, that most troubles us. Similarly, it is the rate at which the magnetic field changes that most affects animals guided by Earth's magnetic field. A reversal of Earth's magnetic field in a year would significantly affect some animals that use the magnetic field (chap. 5). Such a rapid change might be sufficient to trigger some species' extinctions. Fortunately, a reversal of Earth's magnetic field probably takes a few thousand years to occur, during which the field never vanishes. Animals should be able to adapt to such slow changes.

In spite of doubts about all mechanisms so far suggested (such as the examples given above), scientists need to determine if observations, such as those carried out by Ling, link reversals to extinctions. In Ling's case, the statistical correlation between reversals and extinctions is weak. He also did not consider possible artifacts, such as shown in figure 2.5. This cartoon illustrates how depth in a sediment core translates to time. The asterisks and circles represent the locations in a core where two species,

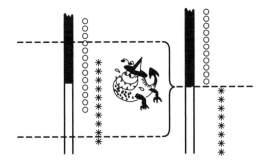

FIGURE 2.5 Erosion or a hiatus can cause an artifact in a sediment core to make it appear as if a magnetic field reversal and species extinction coincide. Asterisks represent the location of fossils with depth for species A and circles for species B. Reverse magnetic field polarity is indicated by black and normal polarity by white in the columns immediately to the left of the circles. In this cartoon, provided by Charles Barton and redrawn by Beth Tully, sediment-eating monsters remove the sediment between the dotted lines in the core on the left to produce the results shown on the right.

A (asterisk) and B (circle), are found. Species A went extinct and is not found in the upper (younger) part of the sediment core. In contrast, fossil evidence of species B extends from the core's surface at the bottom of the ocean, to some depth, before it disappears. A reversal of Earth's magnetic field is recorded in this core and is indicated by a change in black to white in the column to the left of the species "data." The "data" on the left side of figure 2.5 are supposed to represent the situation when sediments were deposited at a constant rate, while the "data" on the right side reflect a hiatus in sediment deposition or the removal of the sediment (between the two dotted lines on the left) by erosion (or sediment-gobbling monsters). As shown on the right side of figure 2.5, the extinction of species A and the reversal appear to have occurred at the same time, even though they did not: a bias has been produced linking extinctions to reversals. I suspect Ling's analysis reflects such a bias or is a consequence of a statistical fluke.

In the late 1970s, Bill Lowrie from Switzerland and Walter Alvarez from California were investigating the magnetization of rocks from the Italian Gubbio formation. This rock formation is used to define one of the most famous geological boundaries—that separating the Cretaceous from the Cenozoic. This boundary represents a change in time when animals, such as dinosaurs, existed and were recorded as fossils in rocks below the boundary to a time (above the boundary) when they were not.[43] Alvarez and Lowrie were interested in the possibility that a magnetic field reversal had triggered the world's most famous mass extinction 65 million years ago. After carrying out careful studies, they demonstrated that no magnetic field reversal occurred at this time. As an aside, during their work in Italy, Walter Alvarez found some fine grain white material just above the boundary. He was curious as to what it was. He asked his father, the Nobel Prize–winning physicist Luis Alvarez (1911–1998), to analyze it. The material turned out to contain large amounts of iridium, an element common to many meteorites and asteroids, but relatively uncommon on Earth.[44] This led the father-and-son Alvarez team (and others) to propose that this mass extinction was caused by a meteorite or asteroid impact (discussed further in chapter 6).

In 1990 Phil McFadden and I were asked to write a review paper on magnetic reversals for the journal *Science*. Based on evidence, such as given above, we concluded (among other things) that there was not a convincing case for magnetic field reversals having caused extinctions. Since then, scientists appear to have lost interest in the subject, even though discussions of it by non-specialists often appear on the Internet.

CHAPTER THREE

Earth's Internal Composition and the Origin of Earth's Magnetic Field

Determining Earth's internal structure and composition is a difficult and ongoing task, made more complicated because Earth's internal properties cannot be uniquely determined from measurements made at its surface. Nevertheless, we need to estimate Earth's internal structure and properties to determine where our main magnetic field originates. This chapter begins by describing how this estimation is achieved.

My wife used to lift and shake Christmas presents to guess what was inside. Although this reduced the uncertainty about the contents, it did not eliminate the non-uniqueness problem: several possibilities still remained. Similarly, it is not possible to determine where the sources of Earth's magnetic field reside from magnetic measurements made at or above Earth's surface. True, we can determine that 99 percent of Earth's magnetic field originates inside Earth, but mathematical uniqueness problems prohibit us from going further. We need other means to determine the source of Earth's magnetic field. We can reduce some of the uncertainty by learning about the structure and composition of Earth's interior.

Sir Isaac Newton used his theory of gravity to determine that Earth's average density was nearly twice that of rocks at the surface of Earth. He concluded that Earth had a dense core. However, modern-day scientists would point out that Newton's conclusions could be interpreted without requiring a core. The distribution of density in Earth's interior is another example of a non-unique problem. A mathematically acceptable, but not a physically reasonable, explanation is that hollow shells alternate with dense layers to explain Newton's results. For that matter, some folklore and pseudoscience suggest the existence of hollow regions deep within Earth inhabited by strange beings. Nevertheless, the pressure of the overlying rocks does not even permit tiny hollow regions beneath Earth's thin crust. But we did not always know this. As late as 1920, Marshall Gardner advocated that Earth's interior was hollow and contained a small sun, which heated our planet.

Scientists have known for a long time that Earth's interior is hot. Evidence for this in the nineteenth century came from eruptions of volcanic material and from underground measurements of temperature in mines. Both indicate that temperature increases with descent into Earth. It was natural to assume that a hot fluid layer, either liquid or gas, was present in the interior. This assumption is also consistent with Edmond Halley's suggestion that Earth contains a liquid layer, which he believed was necessary to account for temporal changes in Earth's magnetic field (chap. 1).

But how does one test for the possibility of a fluid layer residing inside Earth? Gravitational attraction from the Moon and the Sun produce oceanic tides. They also produce much smaller tides in Earth's solid crust. The size of these "solid" Earth tides would be greater if there were a large liquid layer beneath the crust. The German scientist Ernst von Rebeur-Paschwitz (1861–1895) constructed special pendulums to test this possibility. Serendipity appeared: the recording of seismic waves on these pendulums helped to usher in the science of modern observational seismology. Less than an hour after an earthquake was felt in Tokyo in 1889, seismic waves had traveled through Earth and were recorded on Ernst von Rebeur-Paschwitz's pendulums in Potsdam and Wilhelmshaven.[1]

Today seismology provides us with the best information on Earth's internal structure. However, its birth did not begin with Ernst von Rebeur-Paschwitz's pendulums. The British historian Joseph Needham describes a seismoscope developed by a Chinese philosopher, Chang Heng (referred to as Choko in Japan), in 132 AD.[2] It resembled a wine jar with a diameter around six feet. Eight dragons had been attached to the sides of this "jar." Each dragon's mouth contained a ball. Below them sat eight toads looking upward with their mouths open. When the ground shook in a certain direction, a ball fell out of the dragon's mouth into the uplifted mouth of a toad. This supposedly allowed the Chinese to determine the direction from which the earthquake originated.

Much folklore is associated with earthquakes and the seismic waves they emit. Aristotle postulated that winds within Earth occasionally shook Earth's surface. Many cultures have legends tying earthquakes to animals. One ancient legend has the Shinto god Kashima holding a large magical rock over the head of a giant catfish curled up under the islands of Japan. As long as Kashima is diligent, no earthquakes occur. However, when Kashima lets up his guard, the catfish thrashes about, causing Japan to shake. Similarly, a California Indian legend attributes earthquakes to turtles, which hold California on their backs: the turtles move about when they argue and cause California to shake.

The birth of theoretical seismology arguably began with the French mathematician Siméon Poisson (1781–1840), whose work describes the existence of two types of elastic, or seismic, waves. The fastest waves are called primary, P, or compressional, waves, and the slower are secondary, S, or shear waves. I am not certain why we use so many names to describe these waves, but we do. When one repeatedly pushes on a long metal (elastic) spring (or even better, a Slinky), a series of compressions alternating with rarefactions (regions of expansion) are generated, which travel down the length of the spring as waves. These elastic waves are the P waves, sometimes described as "push-pull waves." They differ from S waves, which can be generated by pushing perpendicularly back and forth on the spring.

In the entryway to our home in Seattle, we have a chandelier that acts as a crude pendulum when an earthquake occurs. When the P wave from an earthquake arrives, the chandelier moves back and forth along a line directed toward the epicenter, the location on Earth's surface directly above the focus (also called the hypocenter) of the earthquake. In contrast, the chandelier moves back and forth along a line perpendicular to the direction of the epicenter when a S wave arrives. Although our chandelier's motion is not a particularly good seismometer, most modern-day seismometers make use of pendulums.[3]

Names can affect our perception of whether we understand a concept. Some of you may be struggling with seismic waves at the moment, even though you understand them well in a different context. P waves are ordinary sound waves in air. They are most commonly called acoustic waves in the oceans. We refer to them as compressional, or P, waves in Earth, where their frequencies are lower than we can hear (below 20 Hz). S waves vanish in fluids, such as the ocean or atmosphere, because fluids lack rigidity and cannot transmit shear.

The use of seismic waves to determine the internal structure of Earth is also familiar in a different context. A sonogram uses ultrasound, which has frequencies above our audible range (above 20 kHz) to image a fetus in a pregnant woman. In a similar way, seismologists have been using waves from earthquakes to image Earth's interior for more than a hundred years.

Almost a century ago, Beno Gutenberg (1889–1960), a pioneer of seismology, produced a remarkably accurate estimate for the depth of Earth's core. Gutenberg's father owned a soap factory in Darmstadt, Germany, and he hoped his son would eventually take over the family business. Instead, Gutenberg chose to become a student of Emil Wiechert (1861–1928), who had predicted (from non-seismological data) in 1879 that Earth had a central core beginning at a depth of 1,400 km. Nevertheless, the discovery of the core is usually attributed to Richard Oldham (1858–1936), who in 1906 was the first to use seismic data (P waves) to locate the top of the core at a depth near 3,900 km. In 1912 Gutenberg used seismic travel time data to show that the core began sharply at a depth of 2,900 km (near 1,800 miles), a value still used in many modern-day textbooks.[4]

A travel time refers to the time it takes for a P or S wave to travel from a source, an earthquake or explosion, to a receiver, a seismometer. Gutenberg used travel time data from many earthquakes to find the P wave speed dropped dramatically at the top of the core. Although he also found no evidence that S waves traveled through the core, he was reluctant to suggest that Earth's core was liquid. He worried that S waves might actually travel through a mushy core in which they were damped too much to be detected.

World War I interrupted Gutenberg's seismology work. After being injured in the head by a grenade in 1914 (his helmet apparently saved his life), he worked as a meteorologist to help predict the probability that poisonous gases released by the German army might be carried backward toward them by the wind. After Germany lost the war, Gutenberg earned most of his money by working in his father's soap factory from 1918 to 1930. He also commuted to the University of Frankfurt, where he held the equivalent of a low-paying part-time instructor position. He continued to work in seismology in what little spare time he had. He accepted a full professorship at the California Institute of Technology (Caltech) in 1930. This big leap in title, prestige, and salary finally provided Gutenberg and his wife with enough money to purchase non-essentials. Once in California, he used data from hundreds of earthquakes to construct the first detailed study of the internal seismic structure of the entire Earth.

Unknown to Gutenberg, Harold Jeffreys and his once-student Keith Bullen were working on the same problem. In 1940 Jeffreys and Bullen nearly contemporaneously produced a model similar to Gutenberg's for Earth's internal structure. Jeffreys concentrated on the seismic data, while Bullen developed density models for Earth's interior.

Jeffreys was strongly influenced by a book by the astronomer George Darwin (Charles Darwin's son) on tides. This led Jeffreys, who had received his PhD in mathematics at Cambridge University, to work on the "figure of Earth." The shape of Earth is not a perfect sphere. Instead, the polar radius is about 21 km (13 miles) smaller than the equatorial radius. Earth's shape is described as a spheroid—a flattened sphere. This shape occurs because Earth deforms during rotation. An outward-directed force associated with rotation, the centrifugal force, causes the equatorial radius to be larger. If you tie a string to a balloon filled with water and swing it around, it will be come flatter because of the centrifugal force. Jeffreys found that Earth was slightly flatter than it should be according to existing theory. The difference was small, less than a percent, but Jeffreys was a careful mathematician who was a genius when it came to statistics. He showed the difference was significant, and he attributed the small deviation from the theoretically expected shape as a residual left over from a time when Earth rotated faster.[5] He reasoned that Earth's internal strength was sufficient to maintain a shape inherited millions of years earlier. This also meant, according to Jeffreys' theory, that Earth could not deform fast enough to permit continental drift (chap. 2). Much later, geophysicists showed the small deviations in Earth's shape did not stem from Earth's paleorotation rate but were a consequence of continental drift. It is interesting how the same observational data can sometimes be explained by completely opposite interpretations.

Jeffreys went further than Gutenberg in his analysis of travel time data to conclude in 1926 that Earth's core was liquid. He later teamed up with Bullen to derive the depths of Earth's major seismic and density discontinuities, places where a significant jump in wave speeds occurs. Their seismological model was similar to that produced by Gutenberg and indicates that Earth is divided into a crust, mantle, and core.

The average oceanic crust has been determined to be about 6 to 7 km thick (overlaid by 4 km of water), and the average continental crust is about 35 km thick (although it can reach depths about twice this under some mountain ranges). A discontinuity (jump) in seismic wave speeds

separates the crust from the mantle. Named after its founder, it is called
the Mohorovicic, Moho, or M-discontinuity.

In the late 1950s, a project to drill to the Moho was initiated whose main
goal was to determine the nature of the transition from crust to mantle.
This project turned into a political disaster. In 1966 Congress took the unusual step of killing the Moho drilling project following a series of allega-
tions of corruption. No one has yet successfully drilled to the Moho. This
may change in the near future. The world's most advanced exploration
ship, the Japanese ship *Chikyu*, may drill through a relatively thin section
of oceanic crust in the western Pacific to obtain samples from the mantle.

Seismology data shows that the mantle begins below the crust and extends to the top of the liquid core at a depth a few kilometers less than
2,900 km. It transmits both P and S waves and consists of solid rock. In
1936 Inge Lehmann found seismic evidence for a solid inner core, which is
now located at a depth of 5,150 km and extends to the center of Earth at
a depth of 6,371 km. (As an aside, the contributions of women to science
are now appreciated, but this was not always the case. To acknowledge the
important role of women in the earth sciences, one of the newest medals
given by the American Geophysical Union, the largest earth science society in the world, is named the Lehmann medal.)

The largest changes in seismic speeds are in the radial direction and so
are the major divisions within Earth. The P and S wave velocities essentially increase with depth until the top of the core is reached, where the P
wave speed dramatically decreases and the S wave vanishes because the
outer core is liquid.

However, similar to the imaging of the brain using X-rays, seismic to-
mographic techniques have shown that Earth's mantle also exhibits seismic speed variation in three dimensions. This is particularly true at the
bottom of the mantle, where the so-called D" (pronounced D double
prime) layer exists. This layer varies in thickness from about 200 to 300 km
and manifests lateral variations in chemistry. This will be important to us
when we consider why the rate of magnetic reversals has varied with time.
Nevertheless, on average, the largest changes in seismic wave speed are in
the radial direction.

* * *

While seismology provides valuable information on Earth's structure, it
provides little direct information on Earth's composition, which is not

easy to obtain. Several different lines of evidence must be incorporated in a consistent manner to obtain a good estimate of Earth's composition. Astronomers, geophysicists, geochemists, mineral physicists, and others provide important input. Scientists generally trust information more from research areas close to those in which they work. However, few scientists are familiar with all the fields that provide data used to estimate Earth's internal composition. This forces us into the uncomfortable position of relying on authority arguments. That is also the approach I use here to provide a brief overview of Earth's composition. I begin by discussing the seemingly unrelated topics of the composition of meteorites and our Sun.

Scientists believe the early solar system began 4.6 billion years ago as a disk of gas and dust rotating around the Sun. Several growth stages were required before embryonic planets were formed. Laboratory experiments indicate that the earliest growth stage involved the electrostatic attraction of grains to form large dust grain aggregates. The next stage is not well understood and is based on imperfect computer models. Somehow the aggregates continued to grow until they became mountain-sized "planetesimals," which then exhibited significant gravitational attraction on nearby matter.[6] Finally, low-velocity collisions of planetesimals led to the formation of the Moon and our planets. In this scenario, the solar system essentially formed from chemically homogeneous (or nearly so) material. This is supported by measurements of the chemical composition of the Sun and from measurements of a primitive class of meteorites.

It took a considerable amount of time to discover the origin of meteorites, because it was thought that objects couldn't fall from space. Isaac Newton did not believe small objects could even exist in interplanetary space. A renowned German biologist, Peter Pallas (1741–1811), published a report in 1772 on a 700 kg (1,540 lb.) iron mass he examined in Siberia. Another German, Ernst Chaldni (1756–1827), was motivated by this report to publish in 1794 the conjecture that meteorites, including Pallas's iron mass, originated in outer space. This hypothesis was not well received. For example, Thomas Jefferson is claimed to have remarked to two Yale scientists, who reported on a meteorite find: "It is easier to believe that Yankee Professors would lie, than stones would fall from heaven."[7] People often call meteors "shooting stars" because of the glow they make as they vaporize in our atmosphere. They are typically smaller than the tip of your little finger. Because meteorites are extraterrestrial samples that survived passing through our atmosphere, they are larger. Most of them come from the asteroid belt between Mars and Jupiter. A small number of meteorites originate from other sources, such as Mars and our Moon. When comets or asteroids impacted Mars, they sometimes ejected material that eventually (sometimes after a few billion years) found its way to Earth as meteorites. More than three dozen of Mars' meteorites have been identified, based on chemical analyses (argon isotopes) of tiny pockets of gas thought to represent Mars' ancient atmosphere.

There are many different classes of meteorites, some of which are named after scientists, such as Pallas and Chaldni, but the major classes are named after the properties of meteorites. The largest class of meteorites is called chondrites, because they contain chondrules—spherical inclusions up to a millimeter in diameter. Chondrites are further divided into subclasses, one of which is called carbonaceous chondrites, which contains only a few percent of all known meteorites. Radiometric dating puts the ages of some of these meteorites at 4.5 billion years old. They contain a high percentage of water and organic compounds (amino acids) that would have been lost if they had been heated or chemically altered after they formed.[8] Because these meteorites are very old and have retained their original chemical identity, they are said to be "primitive." Chemical analyses of the chondrules in these primitive meteorites indicate that they essentially have the same composition as the Sun, apart from the most volatile elements, such as hydrogen and helium. This is consistent with the hypothesis that the solar system formed from the same giant molecular cloud as did the Sun. The composition of these meteorites is thought to be identical to, or at least very close to, the bulk composition of Earth.

The Sun should also have a bulk composition close to the molecular cloud from which the solar system formed. Its composition is determined from spectroscopy, which utilizes the observation that light emitted from a chemical element has unique wavelengths, which can be used to identify the element. The relative amounts of these elements can change with time. The Sun is powered by the fusion of hydrogen atoms into helium.[9] This fusion, which occurs in the core of the Sun, also occurs during a hydrogen bomb explosion. (Fusion, which involves the squeezing of atoms of less dense elements together to form heavier ones, is the opposite of fission, in which heavier elements are split apart to produce lighter ones. Fission also releases a considerable amount of energy, as evidenced by the atomic bomb.)

All the natural chemical elements are present in our Sun. Hydrogen is the lightest and most common element: it has one proton around which a single electron orbits. Helium, which is continually being made by the fusion of hydrogen, is the next most common element.[10] Most of the heaviest elements are rare in the Sun. For example, uranium, which is 238 times heavier than hydrogen, is only one-trillionth as abundant as hydrogen in the Sun. Nuclear chemists and physicists have carried out experiments and theory showing that the heaviest elements in the Sun could not have been produced there. Instead, they must represent residual star material, which was produced some time after the birth of the universe 13.7 billion years ago and the formation of the Sun 4.6 billion years ago.[11] The pressures and temperatures in the Sun are not sufficient to produce the heavier elements observed in the solar spectra. More massive stars are required to synthesize these. Indeed, only a supernova can produce elements heavier than iron. The most common type of supernova is produced by a gigantic explosion near the end of the normal lifetime of a massive star (more than eight times the mass of our Sun). It produces elements heavier than iron and then disperses them into interstellar space. The tin in supermarket cans and the nickel in our five-cent coins originated in a supernova. These familiar substances represent recycled star material. They were introduced into the giant gas disk that makes up our solar system before the Sun was born. Solar fusion processes did not create them.

The information gained from the Sun and meteorites provides a good initial estimate for the chemical composition of Earth. But this must be tested to determine whether it is compatible with information scientists have obtained on Earth's properties. One scientist who set out to do this was Francis Birch (1903–1992), a geophysicist who spent much of his career at Harvard. He agreed with the prevailing consensus at the time that Earth had undergone chemical differentiation: when Earth formed, the heavier material settled toward its center. Birch assumed that the core was made primarily of iron, because scientists have identified a class of meteorites, called irons, which are also believed to have formed by chemical differentiation. An iron-rich core was also useful (but not required) in explaining Earth's density. Finally, an iron-rich core would be a good conductor of electricity, a necessary requirement for explaining the origin of Earth's magnetic field (as we will see). Birch concluded that this meant the mantle was made of more common rock-forming minerals, as found in chondritic meteorites. But he needed to test these conjectures. Birch measured the properties of rocks under elevated temperatures and pressures to gain insight into Earth's composition. However, the instrument used by Birch could not achieve pressures much beyond the depth of Earth's crust. A different approach was needed to estimate the composition of the mantle and core.

Birch made many measurements of the chemical and physical properties of minerals and rocks (including seismic wave speeds). He used these measurements in the 1950s when he published several important papers describing Earth's internal composition. In particular, Birch found an important equation relating the seismic wave speed of a mineral to its composition and density. He was also familiar with earlier theoretical work showing how the density of minerals increased with pressure (depth) in Earth. Putting all this together, he estimated Earth's internal composition and concluded that Earth's mantle consists of silicate and oxide minerals, similar to some minerals found in Earth's crust and in chondritic meteorites. Birch attributed the increase in seismic wave speed with depth in Earth's mantle to changes in the properties of these minerals as they were compressed and to a small change in chemistry (a slight increase in the iron content) with depth. The latter was consistent with the notion that chemical differentiation had occurred.

Birch also concluded that Earth's inner core was nearly pure solid iron. The outer core contained about 5 percent nickel alloyed with iron (as found in iron meteorites) and 10 percent of unknown elements with density less than iron. The latter was needed to bring seismic velocity estimates for Earth's core into agreement with the measurements of samples made in the laboratory. Even in the twenty-first century, there is no agreement on which of the lighter elements are present in the core, although oxygen or sulfur are supported by many geochemists. In one classic paper dealing with some of the composition of Earth's core,[12] Birch wrote:

> Unwary readers should take warning that ordinary language undergoes modification to a high-pressure form when applied to the interior of the Earth. A few examples of equivalents follow:

High Pressure Form	Ordinary Meaning
Certain	Dubious
Undoubtedly	Perhaps
Positive proof	Vague suggestion
Unanswerable argument	Trivial objection
Pure iron	Uncertain mixture of all the elements

In spite of a considerable amount of excellent work, including the measurements of samples at pressures and temperatures equivalent to that obtained in Earth's core, our understanding of Earth's core's composition is remarkably similar to that given by Birch more than a half a century ago. The heavier iron settled out of the mantle to produce the core within the first 50 million years of Earth's birth. Birch's description of high-pressure terms is still occasionally shown in modern-day scientific talks to emphasize the uncertainties in estimates of Earth's internal composition. However, our view of Earth's mantle has changed, in part due to the efforts of Ted Ringwood (1930–1993), a geochemist and mineral physicist who joined the faculty at the Australian National University (ANU) in 1958.

Ringwood was aware that chemical differentiation was the accepted explanation for Earth's seismic discontinuities, including some within the mantle, which I have not previously mentioned. However, he thought the seismic discontinuities within the mantle might have a different explanation, perhaps reflecting solid-solid phase transitions. Such phase changes can be understood by considering the chemical element carbon. Although graphite and diamond are both made from carbon, they have different crystalline structures. The carbon atoms in diamond are packed closer together than they are in graphite. Graphite and diamond also exhibit different types of atomic bonding. Because of these differences, graphite and diamond exhibit different physical properties. Graphite is very soft, while diamond is the hardest known natural substance. P and S seismic waves travel much faster in diamond than in graphite. If you squeeze graphite enough, you can transform it into diamond. One can put graphite into a high-pressure device and increase the pressure until it undergoes a solid-solid phase transition to diamond. Although high laboratory pressures are used today to produce industrial diamonds, most natural diamonds originate at a depth of more than 160 km (100 miles) in the mantle, where the pressures are sufficient to convert carbon to diamond. Diamonds are brought to Earth's surface by natural processes in Kimberlite pipes, named after a region in South Africa.

The amount of carbon in the mantle is so small that it has a negligible effect on seismic wave speeds. However, the Australian-born Ringwood wondered if the most common mantle minerals underwent similar solid-solid phase changes with pressure. Ringwood arrived at ANU in 1958, after having spent the previous year working in Francis Birch's laboratory at Harvard. At ANU, he set out to test his conjecture that solid-solid phase changes accounted for the sharp jumps in seismic wave speeds observed to occur occasionally in Earth's mantle. Eventually he measured the seismic wave speeds of minerals from xenoliths under mantle pressures.[13] Xenoliths, solid inclusions of mantle material, are carried up in magma to Earth's surface. Xenoliths come from depths above 200 km and provide valuable information on the composition of Earth's uppermost mantle. They primarily consist of olivine and pyroxene, which are made up of the four chemical elements of magnesium, iron, silicon, and oxygen. Ringwood and others were destined to show that more than 95 percent of Earth's entire mantle is made up of these four elements. By using various high-pressure devices, Ringwood was the first to demonstrate that these minerals underwent solid-solid phase changes, similar to those of carbon, as the pressure increased. The seismic wave speeds in the mantle were consistent with those measured in these mineral phases in the laboratory.

Although in the twenty-first century there still remains considerable uncertainty over the details of the mantle composition, including the possibility of minor changes in chemical composition with depth in the mantle, the bulk of the mantle is essentially chemically homogeneous. Seismic wave speeds usually gradually increase with depth in the mantle, reflecting that minerals making up the mantle are compressed as pressure increases. On occasion, particularly between depths of 410 and 670 km, solid-solid phase changes occur, as recorded by relatively sharp changes in seismic wave speeds and density.[14] In honor of Ringwood's discoveries, one of the high-pressure minerals in Earth's mantle is now called ringwoodite. As we shall learn later, these phase changes also provide us with valuable information on the temperatures within Earth.

On a lighter note, for thirty years I was a frequent visitor to ANU, and I got to know Ringwood well before he died of leukemia in 1993. I also got to know well a far less famous scientist at ANU by the name of Ray Crawford. Crawford and Ringwood had very different personalities. Ringwood was an intense and scientifically insightful man, often self-serving, who became one of the world's most famous geochemists. He received many honors and had considerable influence on Australian politicians. He was even on a first-name basis with the Australian prime minister. In contrast, Crawford was best known at ANU as a practical joker. During his travels, Crawford often picked up stationery to be used for his jokes. During the 1970s, Ringwood became involved in the measurements of the lunar samples returned by Apollo astronauts. NASA viewed these as precious samples, acquired at a great financial cost. Any investigator who wanted to make measurements on lunar samples had to convince NASA officials of the worth of their proposed research and to keep the samples in a locked safe when they were not in use. Ringwood was successful in obtaining some of these samples, which he measured to learn about

our Moon's internal composition. One day he received a letter written on
NASA stationery. The letter informed Ringwood that NASA had funded
psychologists to study the effects that stress had on scientists studying lunar samples. Would Ringwood help in this study by sending a vial of his
urine to the American embassy in Canberra on a weekly basis? Ringwood
complied with this request for several weeks before someone in the embassy had the courage to phone him to inquire what the professor wanted
done with the urine samples. Ringwood, who had a poor sense of humor,
was irate. The story doesn't end there. Within a year, Ringwood received
a phone call from Europe informing him he was the recipient of a new
medal. Ringwood had not heard of this medal, and he slammed down the
phone, incorrectly thinking this was another practical joke by Crawford.
Within a few years, Crawford found it advantageous to leave ANU for a
job in New Zealand.[15]

* * *

The properties of Earth's interior not only depend on composition and
pressure; they also depend on temperature. For example, seismic waves
typically travel slower in materials at elevated temperatures. This is just
one reason that scientists want to know how temperature varies with
depth in Earth. Other reasons include that the cooling of Earth helps to
drive plate tectonics and to maintain Earth's magnetic field.

During the nineteenth century, estimates for the cooling of Earth were
used for another purpose: to estimate the age of Earth. Foremost of the
scientists who did this was William Thomson (1824–1907), who was born
in Belfast, Ireland. By the age of sixteen, Thomson had read the work of
Fourier (chap. 1), which involved the understanding of partial differential
equations, a subject most mathematicians are not exposed to until the
third year in an American university. At the age of seventeen, Thomson
published his first paper on the subject of Fourier analysis. He went on
to make many important contributions to thermodynamics and electricity. During the 1850s, he became involved with a project to lay down a
transatlantic submarine cable for telegraph transmission (between Ireland
and Newfoundland). He became the Lord Baron Kelvin of Largs in 1866
because of his contributions to this project. Kelvin is the name of a river
that runs through the town of Largs on the Scottish coast. Most scientists
know of Thomson simply as Lord Kelvin. He first proposed the absolute
temperature scale, sometimes called the Kelvin scale, in 1848. He had a

remarkable track record for being correct and insightful throughout the first half of his career, but an equally remarkable record of being incorrect throughout most of the second half of his career. Like many physicists, he was sure of himself, and near the end of the nineteenth century he said something to the effect that all the important problems in physics had been solved. He considered geologists to be inferior scientists because of their inability to carry out rudimentary mathematical calculations. They failed to understand that his cooling calculations placed an upward limit on Earth's age.

Thomson was not the first to estimate Earth's age from the length of time Earth had cooled from an initial high temperature state. Comte de Buffon was the first to do this. The French naturalist George-Louis Leclerc (1707–1788), who later became Comte de Buffon, is sometimes described as the pioneer of the theory of evolution. He even speculated that apes and humans might have a common ancestry long before Charles Darwin came on the scene. In 1779 Buffon estimated the age of Earth to be 75,000 years by extrapolating from the cooling rate of a small iron sphere he thought resembled Earth. He was condemned by the Catholic Church of France for this estimate, and many of his 36 volumes of *Histoire naturalle, générale et particulière* were burned. The church "knew" Earth was only about 6,000 years old: they used then the same type of arguments many creationists use today.

Almost a century later, Thomson used a mathematical technique (Fourier analysis) to calculate the age of Earth from the rate it cooled from an initial hot state (7,000°C) to its present state. He incorporated the temperatures measured in mines into these calculations. In 1862 Thomson concluded that Earth was about 100 million years old; the range of his acceptable ages fell between 20 and 400 million years. However, by then geologists wanted an even longer life span for Earth.

The early pioneers of geology had trouble accepting the church's 6,000-year age estimates for Earth. James Hutton (1726–1797) was the first to argue that the present was the key to the past. However, Hutton wrote in an obscure manner, and it took another pioneer to articulate and amplify his basic theme. Charles Lyell (1797–1875), a wealthy Scotsman, traveled to London to become a barrister (lawyer). Because his poor eyesight made this profession impossible, he turned to geology. He proposed one could estimate the age of Earth by examining the rates at which erosion and deposition of sedimentary beds occurred. These processes, he posited, occurred in the past at the same rate they were observed to occur in the present. This so-called uniformitarian approach was criticized by some then and by creationists today. Creationists argue that only a catastrophe could explain the flood experienced by Noah. In the nineteenth century, an emotional debate raged between scientists who believed in catastrophes and those who believed in gradualism, another word used for uniformitarianism.

As an aside, modern-day earth scientists accept that some catastrophes affected Earth's geological and biological history. The best-known example occurred 65 million years ago when an asteroid impacted the Yucatán Peninsula. Not only did this impact generate large tsunamis; it also had devastating effects on the atmosphere and climate. More than 50 percent of all species, including dinosaurs, are estimated to have perished relatively shortly after the impact. This opened the door to allow other animals, including mammals, to evolve rapidly.

Charles Darwin (1809–1882), who thought of himself as a geologist and not a biologist, complained that Thomson's 100-million-year Earth age estimate didn't seem long enough. Thomas Huxley, a great advocate of Darwin, also attacked Thomson; they believed his assumptions must be incorrect. A battle between more qualitative scientists (geologists and biologists) and quantitative scientists (mathematicians and physicists) ensued. Even Darwin's son, the astronomer George Darwin, carried out calculations involving the slowing of Earth's rotation rate and adopted a position opposite to his father's: he obtained an age for Earth consistent with Thomson's. In 1899 Thomson revised his estimate for the age of Earth downward to lie between 20 and 40 million years.

Of course, most of this occurred before radioactivity was discovered. Heat is released during fission processes associated with radioactive decay, something that was unknown before the French chemist Henri Becquerel discovered radioactivity in 1896. Once radioactivity was discovered, geologists quickly realized that the assumptions used by Thomson were wrong and that Earth could be much older than he calculated it to be. Because radioactivity heats the interior, Earth takes longer to cool than assumed by Thomson. (It was later found that the presence of radioactive heat sources within Earth was not the primary reason Lord Kelvin was wrong.[16]) Radioactivity can also be used to date Earth, as first suggested in a lecture in 1905 by the British physicist Ernest Rutherford. In 1911 one of the twentieth century's most famous geologists, Arthur Holmes (1890–1965), used the decay of uranium to lead to an estimate of Earth's age at 1.6 billion years.[17] Much later, scientists concluded that Earth was

born 4.54 billion years ago during a time of great turbulence, which also
led to the birth of our Moon.

Even in the twenty-first century, we do not have a good understanding
of the temperatures within Earth. We are also uncertain how hot Earth
was when it was born. For that matter, we are still trying to determine
precisely how Earth and the Moon were born. The Earth-Moon system is
an anomaly in the inner solar system: none of the other inner planets—
Mercury, Venus, and Mars—has even a moderately sized moon. (This
excludes the two tiny moons of Mars, of which Phobos, the 22 km-long
moon, is the largest.)

The Apollo missions to the Moon provided us with a wealth of information to help us produce models for the origin of the Earth-Moon system. Six Apollo missions, between 1968 and 1972, brought back samples
from the Moon. They left behind four seismometers (and a gravity meter
sometimes used as a seismometer), which sent back data until 1977, when
they were turned off. The seismometers recorded seismic waves generated
from moonquakes, meteorite impacts, and from crashes of various Apollo
stages into the Moon. The data indicate that our Moon, which has a radius
of 1,738 km, is divided into a crust, mantle, and core. A 70 km-thick crust
overlies a mantle that extends to the top of the core, which is less than
300 km from the Moon's center. (The precise radius of the lunar core has
not yet been determined.) The mantle consists of rocks nearly identical to
those of Earth's mantle. These and other observations, such as the orbital
dynamics, need to be explained by any viable model for the origin of the
Moon.

I was recently at a party of non-scientists, where I heard someone explain that our Moon had broken away from Earth in ancient times, leaving
behind a scar occupied by the Pacific Ocean. This is not pseudoscience,
but it is old science. Charles Darwin's son, the astronomer George Darwin,
proposed that Earth once rotated so fast it spun off the Moon. A few years
later, geologists claimed they found the ancient scar left behind when the
Moon broke away: it was the Pacific Ocean basin. This became the favored
explanation for the origin of the Moon throughout much of the first half of
the twentieth century. It is now recognized to be incompatible with plate
tectonics and other data. If you like puzzles, as scientists do, you should try
to determine why this model is inconsistent with plate tectonics.[18] Other
models followed and failed. For example, the conjecture that our Moon
formed far away and subsequently was captured by Earth's gravitational

field fails to explain the similarity of the Moon's composition to Earth's mantle and the existence of the Moon's tiny iron-rich core. The model presently preferred is the giant impact model.

This model is based on computer simulations that suggest a Mars-sized planet, named Theia (a Greek goddess), collided with the proto-Earth.[19] During subsequent aggregation of the vapor and debris resulting from this collision, Earth and the Moon were born. The vast majority of the iron went into Earth, which has a much larger gravitational field than our Moon. This model explains the similar chemical composition of Earth's mantle and the Moon and is compatible with other data, such as the orbital dynamics. The energy of motion (kinetic energy) was converted to another form of energy, heat, when Theia impacted early Earth. The collision arguably produced enough heat to melt much of the mantle of the Moon and Earth. Most planetary scientists suggest that both Earth and our Moon had "magma oceans" shortly after they were born: the entire outer sections of Earth and the Moon consisted of molten material to at least a depth of several hundred kilometers.

Because it takes a long time for Earth to cool, its present temperatures depend on Earth's initial (at birth) temperature distribution. Uncertainties involving the depth of the ancient magma oceans and the locations of heat sources within Earth, such as radioactive heat sources, make it difficult for scientists to calculate precisely the initial temperature distribution. If theory and near-surface measurements of heat flow from Earth's interior were all scientists had, our estimates for the temperatures within Earth would be highly uncertain. Fortunately, the existence of phase changes provides valuable information on the present temperatures. A phase change occurs at a particular pressure and temperature; when the pressure is changed, the temperature also changes.

The pressures within Earth can accurately be determined because the increase in density with depth in Earth is well known. Then by using laboratory data for the temperature dependence of phase changes for the materials constituting the mantle, one can estimate the temperatures where phase transitions occur within Earth. For example, the temperature for the 670 km transition is found to be close to 1,700°C. The temperature generally increases with depth throughout Earth and is much higher in the core than at a depth of 670 km.

A different type of phase transition occurs at the inner-outer core boundary near a depth of 5,150 km. This boundary separates solid iron from liquid iron: it is referred to as a solid-liquid phase transition boundary.[20]

You may be puzzled as to why iron occurs as a liquid in the outer core, while it occurs as a solid in the hotter inner core. To understand why this is so, we need to examine more closely the properties of solids and liquids. When a solid is heated, its volume expands. The average distance between adjacent atoms becomes larger. This allows the atoms to move around more, although they are still bonded together to form a nearly rigid lattice. With enough heating, the spacing between atoms becomes large enough that the motion of the atoms breaks down this lattice. The material transforms to a liquid, marking the temperature of a solid-liquid phase change. In contrast to temperature, pressure squeezes the atoms closer together and makes the material more rigid.[21] In a sense, one can think of melting within Earth as a competition between pressure, which is trying to squeeze atoms together to make iron solid, and heat, which is trying to expand the iron to make it liquid. Throughout the outer core, the heat wins this competition and the iron is liquid. Pressure wins in the inner core, where the iron is solid. Although the transition between the inner and outer core can be thought of as either a freezing or melting boundary, most complex materials exhibit a transition between freezing and melting where solids and liquids coexist. We ignore this complexity in our discussion.[22]

Most estimates of the temperature of the phase transition at the inner-outer core boundary fall between 5,000°C and 5,500°C. These are based on measurements made in laboratories of the melting temperature of iron under high pressure. Theoretical calculations are then used to obtain 4,000°C for the estimate of the temperature at the core-mantle boundary. Similarly, theoretical extrapolation from the inner-outer core boundary temperature inward indicates that the temperature at Earth's center is between 5,500°C and 6,000°C (approximately 10,800°F). These are only estimates, and they may be off by several hundreds of degrees. For comparison sake, the temperature at Earth's center is close to that observed for the surface of our Sun.

As Earth cools over time, the boundary of the inner core size increases: more iron freezes because the core becomes cooler. Calculations suggest that the solid inner core began to form between 1 and 3 billion years ago and has been growing ever since.

The temperature difference between the inner and outer core boundaries is sufficient to cause the outer core to be in convection. This convection is similar to that observed for boiling water, and it is the process by which heat is transferred in the outer core. Convection drives Earth's dynamo.[23]

$$* \quad * \quad *$$

Before we discuss dynamo theory, it is useful to investigate the path that led scientists to this theory. The hypotheses explaining the origin of Earth's magnetic field have changed dramatically throughout history. The belief that Earth's magnetism arose from a soul survived the longest. For a while it was also believed that Earth's magnetic field originated from the heavens, perhaps in the neighborhood of Polaris, the North Star.[24] The permanent magnetization hypothesis, advanced by William Gilbert in 1600, remained favored until Joseph Larmor (1857–1942) suggested in 1919 that Earth's magnetic field was produced by a dynamo in Earth's core. However, long after Larmor's suggestion was made, several scientists retained serious doubts about his mechanism. The physics Nobel laureate Blackett carried out experiments to determine whether some unknown property associated with the rotation of matter produced a magnetic field. His experiments indicated this was not the case (chap. 2). He had not been alone. Several theoretical physicists, including Einstein,[25] had thought the Maxwell equations might be modified to include rotation in such a way as to unify the electromagnetic and gravitational forces. This might provide a natural explanation for magnetic fields in large rotating bodies, such as the Sun and Earth. Other novel explanations for the origin of large-scale planetary and solar magnetic fields had been suggested in the previous century (but not discussed here) and discarded.

In the twenty-first century, theorists will tell you the matter has been settled: Earth's magnetic field is produced by dynamo action (to be explained shortly) in Earth's core. They are sure of this, even though no completely satisfactory dynamo model for Earth's magnetic field has ever been produced. Is this another case of history repeating itself? Some of you may have concluded that "anything is possible" or "everything is relative." Scientists would disagree. While we don't always get everything correct and while we may harbor some mistaken notions about dynamo theory, science shows that some things are not possible and everything is not relative. For example, it is no longer possible that Earth's magnetic field originates in the star Polaris. No *Tyrannosaurus rex* inhabits a large cavern beneath Hawaii and so on. Everything is not relative either. Even Einstein was concerned that the name "Theory of Relativity" would lead to misunderstandings. He thought of his theory as one of invariance: matter cannot travel faster than the speed of light in a vacuum. Einstein's theory of gravity may turn out to be imperfect, but it (and Newton's) is remarkably good at predicting natural phenomena. Of course, one must add a caveat for philosophers: it is possible that the "reality" we see is all in the mind. But how many of you would risk your life by jumping off a tall building to test this?

Dynamo theory is not nearly as well developed as gravitational theory or many other theories in science. Nevertheless, it is now sufficiently developed that I know of no respectable geomagnetist who doubts its validity. We will soon see why this is so.

But before we do this, we need to remind ourselves of the properties of Earth's magnetic field that a viable theory needs to explain. First, the vast majority of the magnetic field originates in Earth's interior. Second, the magnetic field has been present for most of Earth's lifetime.[26] Third, the magnetic field is constantly changing in intensity and direction (secular variation). Finally, the magnetic field reverses from time to time. We could add a host of other phenomena to this list, but those four properties of the magnetic field are the most important undisputed ones. As we shall see, dynamo theory can explain all of these and others not listed. For example, the theory can now even explain why the rate of reversals changes with time and why there have been long intervals of time with no reversals (chap. 2).

The above observations allow us to dismiss some of the theoretical explanations widely accepted in past times. Consider the permanent magnetization explanation. Can it explain secular variation and reversals of Earth's magnetic field? Not without making the model hopelessly contrived. To illustrate this, let's try to construct a permanent magnetization hypothesis to explain Earth's magnetic field. All minerals lose their remanent magnetization above their Curie temperatures, which are well below 1,000°C (chap. 2). Because most of Earth is hotter than this, we need to explain how Earth's interior can be magnetized. We do this by conjecturing that an unknown mineral exists only at high pressures and has a Curie temperature higher than the temperature at Earth's center. This allows us to speculate that Earth's lowermost mantle and Earth's inner core are magnetized. We hypothesize a mechanism similar to Halley's (chap. 1) to explain secular variation: Earth's inner core rotates at a different and variable rate from Earth's mantle. (This is not a very far-fetched assumption, because seismologists have observations indicating that Earth's inner core may be rotating slightly slower than Earth's mantle.) Following Halley's lead, the difference in rotation rate can give rise to secular variation. To explain magnetic field reversals, we posit that chemical alteration affects the magnetization over time to produce self-reversals (chap. 2). I could elaborate on this, but I suspect the reader gets the point: too many conjectures of dubious nature are required to retain a permanent magnetization explanation. Although we cannot formally disprove the permanent magnetization hypothesis, we dismiss it as being too contrived.

I often hear many scientists accusing others of supporting contrived models of some phenomenon. Unfortunately, what is viewed as "contrived" by one scientist is sometimes viewed as "obviously correct" by another. This can be illustrated by assessing the probability that some unspecified hypothesis is correct. Suppose one uses a conjecture that has a 1-in-10 chance of being correct. Would you be willing to rule out a hypothesis based on this conjecture as being too contrived? Consider another hypothesis, for which the validity of two conjectures is needed. We assume the conjectures are independent of each other and each has the probability of 1 in 10 of being correct. The probability that both conjectures are valid and the hypothesis is acceptable is obtained by multiplying 1/10 by 1/10. That is, there is only a 1 in a 100 chance that the hypothesis is correct. Each independent conjecture required lowers the probability that the hypothesis is valid. In particular, our permanent magnetization scenario had more than two conjectures, and the likelihood of any one of these conjectures being true is usually much less than 10 percent. Although it would be difficult to provide a quantitative assessment to dismiss the permanent magnetization hypothesis, we suspect the probability of it being valid is close to zero.

We hear arguments similar to the above used in conspiracy theories: "I think it is possible more than one person was involved in the shooting. It would be in the best interest of, say, the Russians if this person were dead. Of course, the CIA was also involved as evidenced by the fact that one of the Russians met with a CIA agent a month before the shooting. There has been a high-level cover-up by the American government." When a person replies that she does not buy into that hypothesis, she is often asked, "Which assumption do you not agree with?" If she does not think the Russians were involved, the advocate of the conspiracy theory will produce arguments that she cannot dismiss this possibility (for example, "it has a 10 percent chance of being true"). (Readers familiar with the assassination of President Kennedy have heard various conspiracy scenarios concerning his death, some of which are close to the hypothetical example just given.) Similarly, hypotheses emerge in science that chain together a string of conjectures, any one of which when considered separately cannot be safely dismissed. Because we cannot always assign probabilities to all the conjectures, it is sometimes difficult to decide on how contrived a particular hypothesis is. This is one reason we insist on testing hypotheses.

There are only two known ways for producing magnetic fields: magnetization and electric currents. I hope the reader is convinced that permanent magnetization cannot exist much below Earth's crust. If so, this means electric currents must produce Earth's main magnetic field. Strong electric currents in the mantle appear unlikely. The minerals in the mantle are semiconductors: they cannot conduct enough electricity to produce Earth's main magnetic field. By default, the electric currents that produce our main field must reside in Earth's iron-rich core. But how does this work?

In 1820 Hans Christian Oersted carried out an experiment to illustrate to some friends that electric currents caused wires to heat. Oersted was puzzled during this experiment when he observed that a needle in a nearby compass moved every time he turned on the current. The compass needle was neither attracted to, nor repelled by, the wire carrying the current. Instead, it aligned itself at right angles to the current. He later published these results without providing any explanation for them. André-Marie Ampère (1775–1836) expanded on these experiments to show that two wires carrying electric currents magnetically interacted. In the twenty-first century, the relationship between the electric current and magnetic field is often described thusly: the curled fingers of the right hand are along the direction of the magnetic field when the extended thumb points in the direction of the electric current flow (see fig. 1.2). The magnetic field is at right angles to the current, and its intensity decreases with distance from the wire, as observed by Oersted.

Michael Faraday built on the work of Oersted and Ampère. Faraday, the son of a blacksmith, began his career as an apprentice bookbinder and went on to become one of Britain's foremost scientists. One of his contributions was the development of a generator: electric current is generated when a coil of wire is moved through a magnetic field. Faraday could not have guessed his process of converting magnetic field energy into electric energy would be used by Larmor to construct the first dynamo explanation for Earth's magnetic field.

Joseph Larmor produced the mechanical analogue for the dynamo shown in figure 3.1. It requires an initial magnetic field of unspecified

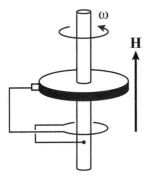

FIGURE 3.1 A disc dynamo. The letter H represents the magnetic field, which points in the direction of the arrow. Rotation of the rod (connected to the disc), which is indicated beneath the rotation symbol ω in the magnetic field, produces an electric current that flows outward in the disc and into the wire. The wire is wound around the axis (rod) of the dynamo in such a way as to reinforce the initial field in the H direction. Figure drawn by Beth Tully.

origin. A metal disc is rotated in that field to induce an electric current. Theory, verified by experiment, shows that the induced current will travel outward from the center of the disc (for the situation shown). This current flows through the brush connected to the rim of the disc, along a wire to a second brush connected to the axis (rod) of the dynamo, and finally up the axis back to the disc to complete the electric circuit. The current in the wire generates a magnetic field. In particular, theory, verified by experiment, indicates that this current produces an upward-directed magnetic field through the center of the coil of wire. That is, in the Larmor model (shown in fig. 3.1), the wire is coiled around the axis in such a way to produce a magnetic field that amplifies the initial one. The more coils in the wire, the larger the magnetic field produced.

Larmor's model is referred to as a disc dynamo model. It has many of the ingredients of twenty-first-century dynamo models. It requires an initial magnetic field, uses rotation (of the disc), and converts mechanical energy into magnetic field energy. Once the dynamo is started, the initial field can be removed—it is no longer an essential ingredient. The magnetic field generated by the dynamo can replace it. Because of this, the dynamo is sometimes referred to as a self-sustaining dynamo. However, we are not generating a magnetic field out of nothing. Mechanical energy must be continually supplied to rotate the disc to induce the electric current. Experiments and theory show that the stronger the field becomes, the more difficult it is to rotate the disc.

Although we don't know the origin of Earth's initial magnetic field, many possibilities have been suggested. One is a thermoelectric effect. Consider a long iron bar that has a higher temperature at one end than the other end. Because the electrons at the hot end have a higher average velocity (they have been heated), more of them move to the cool end than vice versa. This produces a separation of charge that can be used to create an electric current and magnetic field, as was first shown by William Thomson in the nineteenth century. Another possibility is that chemical reactions in the mantle act to produce something akin to a battery. This also can produce a weak electric current and a magnetic field. For that matter, a magnetic field might have been present when Earth formed.

It is not difficult to come up with candidates for the weak initial magnetic field. The problem is how to determine which is the correct one. The initial magnetic field may no longer be present; it is not required for continual dynamo action. There are no rocks to examine for remanent magnetization that go back to the birth of Earth. Therefore, we cannot estimate the properties of the initial magnetic field using paleomagnetism. We can offer many conjectures for what the initial magnetic field was, but no one has been clever enough to find a convincing way to test these conjectures.

Skeptics were more concerned with Larmor's mechanical dynamo analogue than the origin of the initial magnetic field. What constituted the disc, wires, and brushes in Earth's core? No answer could be provided until fluid mechanics was united with electricity and magnetism in the 1930s and 1940s in what is now called magnetohydrodynamics, or MHD. Shortly after MHD theory was developed, dynamo theory suffered a serious setback.

Larmor not only proposed that dynamo theory explained the origin of Earth's main magnetic field; he also claimed that it explained the intense magnetic fields observed on the Sun associated with sunspots. We will talk about sunspots more in chapter 4; it is sufficient here to say that even in the early twentieth century, they were known to have magnetic fields a couple of thousand times the intensity of the magnetic field at Earth's surface. According to Larmor, these strong magnetic fields were produced by dynamo action: a relatively weak initial magnetic field was magnified inside sunspots by mechanical (fluid) motion.

Not so, according to the English mathematician Thomas Cowling (1906–1990), who published a paper in 1933 demonstrating that Larmor's explanation was incorrect. Cowling, who was conspicuous at scientific meetings because he was very tall and had bright red hair, was a skeptic of assertions not backed by rigorous mathematics. He thought something fundamental was wrong with dynamo theory. In 1934 he published a famous theorem, now referred to as Cowling's anti-dynamo theorem.

Although this theorem is still cited today by creationists, who don't like dynamo theory since it allows Earth to be much older than they calculate it to be, Cowling eventually recognized that it was not the general proof he initially thought it was. His theorem only proved that a magnetic field (in Earth's core) symmetric about an axis could not sustain a dynamo. An axially symmetric magnetic field would decay away with time. Modern calculations indicate that such a field in Earth's core would decrease by 63 percent after every 15,000 years or so.[27]

For the next three and a half decades, mathematicians sought a general anti-dynamo theorem. They produced additional conditions under which dynamos could not operate. They showed that Cowling's theorem required both the fluid motions and the magnetic field to be symmetric about an axis. Usually this axis is taken to be Earth's rotation axis. It is the only axis for which such symmetry could conceivably exist because the forces associated with Earth's rotation would make fluid and magnetic symmetry about any other axis impossible. But even symmetric motions about the axis of rotation seem unlikely. If such symmetry existed for the fluid (gas) in our atmosphere, our weather would not change around any given latitude, regardless of whether continents or oceans were present. A hurricane at the equator would have to encircle the entire Earth. Clearly such axial symmetry does not exist in our atmosphere, and similarly it almost certainly doesn't exist in Earth's outer core, where we now think a dynamo operates. The search for a general anti-dynamo theorem ended when mathematicians, particularly Steve Childress and Glen Roberts (working independently) in England, proved in 1970 that such a theorem did not exist. Even more importantly, they provided reasons to believe dynamos should be common in the fluid cores of planets and stars.[28]

I have heard physicists complain that mathematicians have "too big a hold" on dynamo theory, and as a consequence some of the most important problems in the subject get lost in the shuffle. You might be puzzled by such a comment. How can someone get "too big a hold" on a subject? One conceivable way involves how the review process works. Mathematicians and physicists often approach problems in different ways. Mathematicians emphasize proofs, while physicists often use physical insight to get past difficult barriers. A reviewer who is a mathematician might recommend rejection of a manuscript for publication because it contains a conjecture not rigorously proved, while a physicist might recommend rejection because the manuscript doesn't contain sufficiently new science. ("It only dots the i's.")

In reality, it is difficult for a group of scientists to monopolize a subject, and I suspect it rarely happens. Nevertheless, even Hannes Alfvén
(1908–1995), who received the physics Nobel Prize in 1970 for his work
on magnetohydrodynamics, once complained to me over dinner that he
couldn't get his work published in mainstream physics or geophysics journals: "The reviewers don't understand the fundamentals of the new concepts I am proposing." Alfvén, who spoke several languages fluently, was
a rebel. It seemed to me he often intuited the answer and then searched
for the physics to support his conclusion. Often he was correct, but there
were some notable exceptions. He once confessed he had not supported
the big bang origin for our universe. This cosmological model asserts that
our universe expanded from a primordial state to our present state. The
expanding universe model is widely accepted by modern astronomers (in
slightly modified form). Because the big bang seems to imply that our
universe had a beginning, he thought it was a myth designed to support religion. His views were often blunt and unpopular. Alfvén's remarks to me
indicate that even Nobel Prize laureates sometimes feel unappreciated.

Alfvén used a concept called dualism in many of his papers. Dualism
means there are two (or more) different ways to solve a problem. The
investigator tries to pick the easiest one to use. An example is finding
the circumference of a circle. One can measure the circumference, a geometric approach, or one can use an algebraic equation and calculate the
circumference. Scientists working in MHD theory have found it convenient to use the famous Maxwell's equations to convert all electric fields to
magnetic fields. Rather than using the electric current in their equations,
they convert all the currents into equivalent magnetic field descriptions,
which makes solving problems easier. Of course, they can also calculate
the electric currents afterward—if they want to.[29]

Using this approach, theorists discovered an equation central to dynamo theory called the magnetic induction equation. It describes how
the magnetic field changes with time. It shows that a magnetic field can
change through diffusion, which occurs in a manner similar to heat diffusing through a wall.[30] Diffusion causes the magnetic field to decay over
time. Earth's magnetic field would be near zero after 100,000 years—if
only diffusion were occurring.[31] We can safely conclude that our field is
not a remnant of a field formed 4.6 billion years ago when Earth was born.
Such a field would have decayed away long ago.

The magnetic induction equation also indicates that the magnetic field

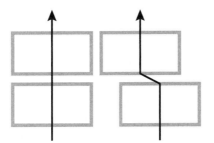

FIGURE 3.2 "Frozen in field" concept. The left side shows a magnetic field line, indicated by an arrow at its end, penetrating two perfect conductors. The right side shows how the field line is distorted by the relative displacement of the lower conductor to the right. A new portion of this line is produced at the boundary of the two conductors. This portion of the field line is a new field—one that is directed nearly parallel to the two conductors. If the conductors on the right had no gap between them (as they would be in the liquid iron case in the core), the new field would only be in the horizontal direction. Figure drawn by Beth Tully.

cannot diffuse through a perfect conductor.[32] This provides a second way a magnetic field can change. When a perfect conductor moves, it carries with it the magnetic field: the field is said to be "frozen into the conductor," a concept developed by Alfvén. Consider what happens when a perfect electrical conductor undergoes shear, as shown in figure 3.2. The relative displacement of the two conductors generates a new magnetic field along their boundaries; the field line shown must remain connected, and it cannot diffuse within the perfect conductors. We have not obtained the new field from nothing: it took energy to displace the conductors.

New magnetic field in Earth is produced when the core fluid does not move uniformly. Although in Earth the conducting fluid carries the magnetic field along, some diffusion also occurs because the iron-rich core is not a perfect conductor. A necessary requirement for a dynamo is that the average production of new magnetic field must be greater than the loss of magnetic field through decay. No magnetic field can be sustained if it decreases faster than it is built up.

While mathematicians were trying to find an anti-dynamo theorem, physicists were trying to develop a viable dynamo model using insight gained from processes such as illustrated in figure 3.2. Were there fluid motions that could produce a new magnetic field fast enough to counter decay?

Walter Elsasser (1904–1991) is often viewed among theorists as the founder of theoretical dynamo theory. Elsasser, a German who found out he was of Jewish descent when he was fifteen, escaped from Nazi Germany to France and then to the United States. He developed the mathematical structure to illustrate how dynamo theory might be successfully developed.

This helped pave the way for others, such as Eugene Parker, a University of Chicago physicist who thought the outer fluid core was not uniformly rotating. When you stir coffee in a mug, all the coffee does not rotate at the same rate: the coffee toward the center of the mug rotates faster than that toward the outside. Parker conjectured that the differential rotation of fluid in the core would produce a new magnetic field parallel to latitudes in the core in a similar way that the horizontal magnetic field is produced in figure 3.2. This field, represented by circular field lines about Earth's rotation axis (called a toroidal magnetic field), could not produce a dynamo according to Cowling's anti-dynamo theorem. Parker needed something more. He assumed Earth's core was convecting. Convection is now known to be the dominant way heat is transferred in Earth's outer core. It is also the dominant way heat is transferred in boiling water: hot water rises and cold water descends.[33] The convection of fluid in the outer core carries along the magnetic field. Because hot liquid moves upward and cold fluid moves downward, a new magnetic field will be produced. The moving fluid is also subject to the Coriolis force. This is a force present because Earth rotates, and it is the force responsible for producing the anti-clockwise rotation of winds in a hurricane in the Northern Hemisphere and the clockwise rotation of winds in a cyclone in the Southern Hemisphere. Similarly, we expect there to be clockwise rotation of core fluid in the Northern Hemisphere during "core storms" (and vice versa in the Southern Hemisphere). Parker showed how the Coriolis force twisted the fluid and magnetic field into closed loops (figure 3.3), which then coalesced to form the core's dipole field. He produced an intuitive picture showing how differential rotation and convection in the outer core might produce a dynamo.

Although the Parker dynamo model is a good teaching aid, it has a serious deficiency. The magnetic field in Parker's dynamo would continue to grow exponentially over time and become ludicrously large in just several hundred thousand years. Parker did not include an important feedback between the magnetic field and the fluid. For mathematical simplicity, he omitted theory showing that the larger a magnetic field becomes, the more it resists fluid motion. We can understand this feedback by making an anal-

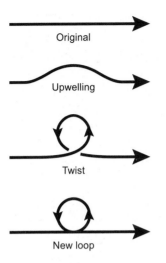

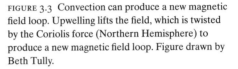

FIGURE 3.3 Convection can produce a new magnetic field loop. Upwelling lifts the field, which is twisted by the Coriolis force (Northern Hemisphere) to produce a new magnetic field loop. Figure drawn by Beth Tully.

ogy between a magnetic field line and a rubber band. When displacement of the conductors in figure 3.2 occurs, the rubber band (field line) is bent and stretched. If the force causing this displacement is removed, the tension in the rubber band will return the conductors to their initial relative positions. (One on top of the other, as shown on the left side of figure 3.2.) A larger magnetic field is represented by more field lines (rubber bands). That is, the larger the magnetic field is, the larger the resistance is. This feedback of the magnetic field on the liquid makes dynamo theory difficult. A dynamo is a process that magnifies an initial magnetic field by the relative movement of an electrically conducting fluid. As the field grows, it alters the fluid flow. The modified fluid flow then alters the magnetic field, which again modifies the movement of the fluid. And so on it goes. Such feedbacks between the magnetic field and the fluid flow make it difficult to solve the Earth dynamo problem. Parker ignored these feedbacks, but others that followed tried to take them into account.

I will fast-forward through a lot of developments in dynamo theory to 1996 when Gary Glatzmaier (then at Los Alamos) and Paul Roberts (at UCLA) produced the first "self-consistent three-dimensional" dynamo for Earth, including one with a solid inner core. Because this dynamo took into account the feedbacks between the fluid motion and the magnetic field generation, it is referred to as a self-consistent dynamo. Its solution required 2,000 hours on a supercomputer. Their dynamo even reversed

FIGURE 3.4 Normal polarity for the Glatzmaier-Roberts's dynamo. The lines shown represent magnetic field lines (white) that mostly exit Earth's core in the Southern Hemisphere and enter (gray) in the Northern Hemisphere. The core is approximately where the spherical bundle of field lines is in the center of the figure. Figure provided by Gary Glatzmaier.

polarity. The magnetic field for normal polarity in this dynamo model is shown in figure 3.4, and the field during a reversal is shown in figure 3.5. Dynamo theory had finally matured to the point where it could be tested using results obtained from paleomagnetism.

The Glatzmaier-Roberts's dynamo explains many of the observations of Earth's magnetic field, including reversals. However, difficulties arose. A year after Glatzmaier and Roberts published their dynamo, the Harvard team of Weijia Kuang and Jeremy Bloxham published a different self-consistent three-dimensional dynamo also compatible with most of the observational evidence. Subsequently, many other dynamo models have been produced that consistently treat the feedback between the core fluid and the magnetic field. The fluid motions producing the magnetic fields in these dynamo models are often dramatically different from each other. This occurs because different theorists use different simplifying assumptions leading to different outcomes. None of these models adequately treats Earth's dynamo.

FIGURE 3.5 The magnetic field during a reversal. The magnetic field during a reversal in this Glatzmaier-Roberts's computer simulation shows a strong nondipole field component, with several places where the lines are vertical entering (gray) or exiting (white) the core. The scale size is the same as in figure 3.4. Figure provided by Gary Glatzmaier.

A particular shortcoming of all modern dynamos is that they cannot treat turbulence in a self-consistent manner. We can gain some appreciation of turbulence by first considering water flowing smoothly, not turbulently, in a canal. It is a simple matter to predict the velocity of water several tens of meters downstream: it is essentially the same as you observe at your location. This is not the case for the white-water rivers that kayakers enjoy navigating. One cannot accurately predict the turbulent flow of a white-water river even a short distance downstream from a location at which the velocity was measured. Using velocity estimates of the core fluid obtained from secular variation, theorists conclude that Earth's outer core is turbulent. The physics of ordinary turbulence is poorly understood, and the physics of turbulence in a magnetohydrodynamic (MHD) system is even more difficult to understand. All self-consistent dynamo models assume there is no turbulence. There are also dynamo models that take turbulence into account, but they are not self-consistent: they don't properly treat the feedback between the magnetic field and the fluid motion and vice versa. Even with supercomputers, we are still far from producing a realistic dynamo model for Earth.

Let's summarize how we think Earth's magnetic field originates without using MHD jargon. There are only two known mechanisms for generating a magnetic field. The first is magnetization, and the second is electric currents. Because Earth's core is too hot to be permanently magnetized, electric currents in the electrically conducting core are thought to produce Earth's main magnetic field. The dynamo requires an initial magnetic field, the origin of which is not well known. Differential rotation and convection induce electric currents in the iron-rich core in a manner to amplify the initial magnetic field. Once the dynamo is started, the initial magnetic field is no longer required. The dynamo is sustained by outer core convection.

Why do we believe this picture is valid even though we do not have an adequate dynamo model for Earth? We have many dynamo models, including ones that produce magnetic field reversals. There is no scientific objection to dynamo theory; we are just not up to the task of solving the difficult dynamo problem that has parameters appropriate for Earth. MHD and many aspects of dynamo theory have been successfully tested. Some tests are even carried out on dynamos in scientific laboratories, where liquid sodium, a metal that conducts electricity well, is substituted for liquid iron (for theoretical reasons I will not get into). I am confident that a hundred years from now, we will still attribute the origin of Earth's magnetic field to a dynamo. However, I would be surprised if the details of the dynamo looked like any of the existing models. Of course, given the history of science, I could be wrong.

* * *

One of the tests of dynamo theory is whether it can produce phenomena we observe, such as magnetic field reversals. As we have seen, modern theory can do this. But this was not always so. Less than two decades ago, several scientists thought sources external to Earth's core might have triggered reversals. When meteorites or asteroids impact Earth, they transfer most of their energy of motion (kinetic energy) into heat. But some of it also goes into the kinetic energy of molten rock fragments dislodged on impact. These fragments rapidly cool in Earth's atmosphere to form glassy objects called tektites. In the 1960s some geologists thought tektites occurred more often than by chance when magnetic field reversals occurred. They wondered if reversals were triggered by a shock wave from a meteorite altering the fluid motions in Earth's core.

Richard Muller and Donald Morris at the Lawrence Berkeley Laboratory of the University of California revived the impact hypothesis in 1989, but with a twist. A large impact, they argued, would loft dust into the atmosphere, cool Earth, and increase the size of the polar ice caps. This would speed up Earth's rotation, much like an ice-skater spins faster by pulling her arms to her body. Muller illustrated the effects that a change in Earth's rotation would have on the outer core by showing how a raw egg spins differently from a hard-boiled egg. The liquid outer core would respond like the raw egg and a reversal would occur, he argued. However, Muller did not provide any mathematics to show how this intuitive picture produced a reversal.

While these speculations infuriated many scientists, none was more upset than Dave Loper at Florida State University. Loper, an applied mathematician, had spent much of his career studying processes acting in Earth's interior. He once asked me something like, "Why didn't these physicists from Berkeley at least glance at a few of the many excellent papers dealing with thermodynamics and dynamics of Earth's interior?" I had no answer. Loper knew that changes in mass on Earth's surface, such as moving water from the ocean to the ice caps, would affect Earth's rotation rate. However, he also knew this would have a minuscule effect on Earth's dynamo processes. At one national scientific meeting, I was asked to convene a special session at night to moderate an argument that had erupted earlier in the day between Loper and Muller. At one point while Muller was speaking at this special session (which was attended by reporters as well as scientists), Loper became so upset he jumped up and wrote several equations on a board to falsify Muller's argument. Unfortunately, while Loper thought he was writing these on a white board, it turned out he had written them on a screen used for the projections of overheads. Although Loper's arguments seemed valid to me, the audience reaction was first of alarm, followed by laughter, which was amplified by Loper's contrite demeanor after he realized what he had done. This event also illustrates the different emotional responses to science published in peer-reviewed journals. Some scientists who attended this session thought the Muller and Morris paper had stimulated research in the area of magnetic reversals, while others felt it served as a distraction and disservice to sound science.

Later I coauthored a review paper with my Australian colleague Phil McFadden for the journal *Science*, in which we dismissed the speculation that impacts cause reversals. Geologists had revisited the rock formations containing tektites to show there was no convincing statistical evidence for reversals and tektites occurring at the same time. The large asteroid collision 65 million years ago that caused the demise of the dinosaurs was not associated with a reversal or a magnetic field excursion. The physical mechanism for producing the mechanism was also inadequate, as had been correctly articulated by Loper.

Other speculations for producing reversals by processes acting outside of Earth's core meet a similar fate: when analyzed closely, they fail. I am not aware of any new speculations of this sort being advanced during the twenty-first century, probably because reversals can be naturally explained within the context of dynamo theory. The English scientist Dave Gubbins, who received the 2004 Fleming Medal, was asked in a 2008 interview published in *Nature*, "Why does the magnetic field reverse?" His succinct response was, "Because it can."[34] I sometimes reply to such a question with: "Reversals occur by the convection processes producing the dynamo. There is no need for a mechanism outside of Earth's core to produce a reversal."

During the 1990s, Gubbins visited the University of Washington to give us a talk on his research. One day I noticed a faculty member on the phone trying to arrange for someone from a local dive shop to go scuba diving with Gubbins, who was waiting nearby for a reply. The faculty member looked up to see me and said, "Why don't you go with Ron—he's a scuba diver." While I am a friend of Gubbins, I had several years earlier turned into a warm-water diver. I had given up going in the cold waters of Puget Sound. Of course, "cold" is a relative term, and what is cold to me was not particularly cold to Gubbins, who often dived off the coast of England. The next day found the two of us wading in thick wet suits into the waters of Puget Sound to investigate a wreck. After about thirty minutes of diving, I gave the signal to Gubbins to ascend. At that point he swam over to me and looked at my air gauge. It indicated my tank was a little over half full. Gubbins simply shook his head and continued on with the dive. Unfortunately for me, Gubbins is an excellent scuba diver who went through the air in his scuba tank rather slowly. Eventually he took pity on me, perhaps because the color my face had taken on a blue hue, and we ascended.

Afterward we made some comparisons between the ocean and the outer core. There can be a sharp decrease in temperature called the thermocline in the ocean. However, on a few occasions (alas, not on the day of our dive), when a temperature inversion occurs, the surface water will be colder than the underlying water. Because warm water is less dense, it is buoyant and will rise, if given the chance. The density of ocean water is also affected by how much salt is dissolved in it: the more salt, the higher the density. If the underlying layer has the same temperature but lower salinity, it will be buoyant and try to rise. When arctic water freezes, it releases salt into the ocean. This is a stable situation because the salty oceanic water is even denser than fresh water and the ice floats. But the opposite situation occurs in Earth's core. As Earth's inner core grows over time by freezing, it releases less dense fluid containing oxygen or sulfur into the outer core. This buoyant fluid rises at a rate near a millimeter per second[35] to produce convection in the outer core. This chemical convection is thought by many theorists to be even more important in the outer core than convection driven by differences in temperature (thermal convection).

Since chemical convection began only after Earth had an inner core about 2 billion years ago, scientists are trying to determine if there had been a fundamental change in the dynamo since then. Before an inner core formed, only thermal convection powered the dynamo. Unfortunately at this time, paleomagnetic data are inadequate to determine if the properties of the dynamo changed after a solid inner core formed.

However, there are paleomagnetic data suggesting that the rate of reversals can be affected by changes over time in the D" layer at the base of the mantle and possibly by the presence of the inner core. In 1988 Phil McFadden and I speculated that the pattern of heat flow out of the core changed slowly with time and affected the rate at which reversals occurred. The pattern of heat flow, we reasoned, would affect certain symmetry characteristics of the magnetic field. The magnetic field can be divided into a component symmetric with respect to the equator and one anti-symmetric to the equator. Examples of symmetric and anti-symmetric components are respectively the axial quadrupole field and an axial dipole field, as shown in figure 2.4. We thought that the probability of reversal might be higher when the symmetric part of the field was relatively greater than the anti-symmetric part. If this speculation is true, it should be manifested in all the temporal variations of the magnetic field—not only those associated with reversals. We teamed up with our Australian colleague Mike McElhinny to test these ideas using paleomagnetic data. Although the test was positive, it did not convince several geomagnetists, who thought the idea would not hold up once dynamo theory matured to the point it could be used to test our idea.

After Glatzmaier and Roberts produced the first self-consistent three-dimensional dynamo model for Earth, they teamed up with a paleomagnetist, Rob Coe, to show how the rate of reversals could be controlled by the manner that heat was conducted through the D" layer into the mantle. The spatial pattern of heat flow out of the mantle affected the rate of reversals. For example, fewer reversals occurred in their dynamo simulations when heat flow out of the core was preferentially concentrated in the equatorial region rather than at mid-latitudes. Although changes in D" are slow, reflecting the slow creep of mantle material at about 10 million times less than the convection rate in Earth's outer core, they appear sufficiently fast enough to explain the slow changes of the magnetic field reversal rate over time (chap. 2). Nevertheless, these scientists had not directly tested our symmetry speculations. This had to wait until 2006 when Rob Coe and Gary Glatzmaier (both now at the University of California at Santa Cruz) showed in an article in *Geophysical Research Letters* that our speculations are compatible with modern dynamo theory. While we are encouraged that our ideas are correct by such analyses, it remains unknown whether the predictions made about reversals based on the Glatzmaier and Roberts's dynamo will be sustained when even more realistic dynamo models are produced.

In 1993 the English mathematicians Rainer Hollerbach and Chris Jones carried out calculations to suggest the inner core acts to stabilize the polarity of Earth's magnetic field. After a magnetic reversal occurs in the outer core, it diffuses into the inner core to replace the previous field. The inner core thus provides some inertia to maintain the present polarity of the field produced in the outer core. The dynamo retains some memory: it takes a few thousand years for diffusion of the remnant magnetic field out of the inner core to occur. This is a simplification. In reality, a complicated feedback exists between the fields in the outer and inner core, requiring detailed computer analyses. Computer simulations suggest that the axial dipole field is relatively stronger and reversals are less frequent when the inner core is moderately large. Reversals may have been more common in Earth's early history; the inner core has grown over time as Earth has cooled.

Dave Gubbins suggests that the feedback between the field in the inner core and the outer core might explain magnetic excursions, often viewed as aborted reversals (chap. 2). Because the field reverses polarity in the outer core before it does in the inner core, the polarity of the inner core encourages the field in the outer core to return to its initial state. When this happens, an excursion occurs.

Clearly, to quote a scientific cliché, more research is needed.

* * *

Earth is not the only planet to have a magnetic field produced by a dynamo. Moreover, it is not even the only planet in our solar system that has experienced magnetic field reversals.

In 1999 I received a phone call from NASA headquarters. Would I fly back to the East Coast to participate in a press conference dealing with recent results obtained from measurements of Mars's magnetic field? I explained I was not part of that mission and did not know what the scientists had found. No problem: they would electronically transmit an article to be published in the journal *Science* that I could read on the plane. They wanted an expert on Earth's magnetism to be able to respond to reporters' questions on the importance of the new results. The first two authors of the thirteen-authored *Science* paper, Mario Acuña and Jack Connerney, would present the major findings of the mission. I agreed to fly back to Washington, DC, as this would be a great opportunity to be updated on Mars's magnetic field, which was essentially thought to be non-existent before the 1999 Mars Global Surveyor Mission.

As an aside, agencies such as NASA like to see results of their efforts published in newspapers and magazines and shown on TV. In this competitive world, they believe it helps them secure funds from the government. Although reporters attend these conferences and more read the press releases, many appear to obtain their information from reading two journals: the American-published journal *Science* and the British-published journal *Nature*. These journals compete with each other to produce cutting-edge newsworthy papers. The formats of the journals are similar: most papers are four pages in length, including references, and about a half-dozen short commentary articles are published simultaneously to emphasize the relevance of the papers to scientists (and reporters) not in the research specialty. *Nature* and *Science* have become two of the most prestigious journals in the world. They publish short articles quickly, have a very high rejection rate for manuscripts submitted, and are well read and cited. Their articles also often contain a considerable amount of speculation. One of my geomagnetic colleagues, who had recently had a manuscript rejected by *Nature*, complained to me, "They rejected the manuscript on the grounds it would probably turn out to be wrong. Everyone knows almost everything published in *Nature* and *Science* is probably wrong." He has a point, because most science at the frontier of any subject is modified within a few years. The pressure to publish in prestigious journals is only increasing. In 2006 South Korea joined China and Pakistan to offer cash rewards to scientists who publish in *Nature* or *Science*. The rewards range from a thousand to a few tens of thousands of U.S. dollars, depending on the country and institution involved. *Nature* and *Science* do not applaud this practice. The editors of *Nature* even wrote an editorial in June 2006 suggesting that this practice could increase attempts at scientific fraud.

Besides Earth and our Sun, Jupiter, Saturn, Uranus, and Neptune have active dynamos. Most planetary scientists think Mercury also has one. (A few scientists argue that Mercury's weak magnetic field is derived from rocks magnetized in an ancient dynamo.) Even one of Jupiter's moons, Ganymede, appears to have an active dynamo. The remanent magnetization in crustal rocks produces the weak magnetic field of the Moon and Mars, which no longer have active dynamos. Venus does not have a dynamo, probably because it does not rotate fast enough. Venus only rotates (in a retrograde direction) once every 243 Earth days, which many planetary scientists believe is insufficient to produce a dynamo.[36]

Jupiter has a dipolar magnetic field 2,000 times larger than Earth's. It is the largest in our solar system—and that includes our Sun.[37] Saturn also has a large dipole field, which is 600 times as large as Earth's. Even Neptune and Uranus have magnetic fields substantially larger than Earth's. With the exception of Ganymede and possibly Mercury, Earth has the smallest known global magnetic field produced from a dynamo in our solar system.

It is difficult to know whether dynamos in other planets occasionally reverse polarity. As far as we know, a dynamo does not have to ever reverse. Earth has two mega-states, one in which reversals occur and one in which they do not (chap. 2). Some scientists have speculated that Neptune and Uranus are currently undergoing a reversal, because both planets have dipole fields tilted at large angles with respect to their rotation axes. But there are alternative explanations for these large dipole tilts.[38] In contrast, Jupiter's and Saturn's magnetic dipoles are relatively close to their rotation axes, and they do not appear to be in the process of reversing polarity.

Dynamos seem to be common throughout the universe. Stars exhibit a wide range of magnetic fields. Some stars even have surface magnetic field strengths 10^{15} (a million billion) times larger than the average magnetic field at the Sun's surface. These stars, named magnetars, are exotic forms of neutron stars—remnants of supernova consisting of densely packed neutrons.[39] Dynamos are the primary, but not the only, mechanism for producing magnetic fields in stars. Probably there are also some important differences in the nature of dynamos in different astronomical bodies. Dynamos are even the favored mechanism for producing the magnetic fields observed in spiral galaxies, including our galaxy, the Milky Way. The average magnetic field in the Milky Way is estimated to be around 1/50,000 of the magnetic field at the surface of Earth. The magnetic field lies in the plane of our galaxy, as it does for other spiral galaxies, and it is largest toward the center of the galaxy.

These results have some important lessons for earth scientists. Perhaps the most important is that dynamos are remarkably robust features in our solar system and the universe beyond. The data indicate that only two or three conditions are needed for stars and planets to have dynamos. There must be a large fluid region of electrically conducting material. Second, this region must be convecting. Finally, I suspect there must be sufficient rotation of the object.[40]

Let's return to Mars. The data obtained in the 1999 paper in *Science* came about because of a NASA "failure." A mistake led to greater braking of the Mars Global Survey Orbiter, forcing it into a lower orbit than intended. This fortunate error led to some startling magnetic measurements recorded by a magnetometer on board and relayed to Earth. Mars exhibits magnetic anomalies in the form of stripes, reminiscent of Earth (chap. 2). These anomalies are the largest in our solar system: they are ten to twenty times larger than any observed for Earth (as would be measured at a 100 km elevation above the anomaly). The anomalies come from very strongly magnetized rocks close to the surface of Mars. Even more surprising, the anomalies are mostly confined to the Southern Hemisphere and exhibit east–west lineation. The authors of the 1999 *Science* paper concluded that Mars once had a dynamo that reversed polarity to produce the stripes. By clever geological detective work, they estimated that the dynamo turned off around 4 billion years ago. They also speculated that plate tectonics once occurred on Mars, but this remains controversial.[41]

In 2002 a student working for Joe Kirschvink at Caltech, Ben Weiss (now on the faculty at MIT), and his colleagues made paleomagnetic intensity (appendix) measurements on a meteorite recovered on Earth that came from Mars to conclude that the strength of Mars's paleomagnetic field was similar to the intensity of Earth's present field. But today Mars has no dynamo-driven magnetic field, while Earth does. Presumably Mars had cooled sufficiently during the first half billion years of its existence that convection could no longer power a dynamo. Because Earth's core is larger than that of Mars, it cools slower. A long time in the future, our dynamo will also shut down.

Mario Acuña, the lead author of the *Science* paper, was asked in the 1999 NASA press conference whether magnetic measurements had consequences for the search for primitive life on Mars. He responded that the atmosphere would have been different when Mars had a magnetic field. He explained, as is further discussed in the next chapter, that the magnetic field would have protected Mars's atmosphere from solar radiation. After the main magnetic field ceased to exist, the composition and size of Mars's atmosphere was altered. Acuña's speculation is consistent with some recent interpretations of data from the Mars Exploration Rover Mission suggesting that Mars was warmer and wetter when it had a magnetic field.

Scientists are not worried Earth's dynamo will cease to exist in the near future; it should hang around for at least many hundreds of millions of years. However, there are still potential changes in our magnetic field that might have serious consequences for us. Most dramatic of these are associated with reversals of Earth's magnetic field, during which the intensity substantially drops. One possible consequence of this was discussed in chapter 2, and others are discussed later in this book.

CHAPTER FOUR

The Sun-Earth Connection

Although most people are familiar with the northern lights and have heard of power outages caused by magnetic storms, few know their beginnings can be traced back to our Sun. Solar phenomena, including magnetic field activity, act as seeds that sprout and extend their tentacles 93 million miles to produce magnetic storms and substorms in Earth's upper atmosphere. To begin this journey, our discussion focuses on the character of the diverse magnetic fields on the Sun, which requires knowing something about the Sun's structure. As with Earth, the best information on the solar structure comes from seismology.

"Where is that #*&#* student who located the earthquake under San Francisco Bay?" bellowed Perry Byerly (1897–1978) on a Monday morning in the early 1960s. Byerly was an eminent seismologist who supervised the University of California at Berkeley's network of seismograph stations for thirty-eight years. During the previous week, he had (unknown to me) consulted with BART (Bay Area Rapid Transit) engineers, who were in the planning stages of constructing a rapid rail system in a tunnel under the San Francisco Bay. Byerly had assured them that earthquakes did not occur under the bay. As a student employed by Byerly, I wondered if I had incorrectly located the small earthquake that had occurred on Sunday— a distinct possibility because this was before computers were routinely used to locate earthquakes. Instead, we used pen and paper to determine the epicenter and earthquake magnitude quickly so that reporters could be provided with this information. If I had mislocated the earthquake, I would have been in deep trouble. Fortunately, I had not done so and I retained my research assistant position.

Only a couple of years earlier, in 1960, seismologists had detected the natural frequencies of Earth, when it rang like a bell for several weeks after being triggered by a large earthquake in Chile. These natural frequencies (resonances), which are detected by sensitive seismic instruments, provide valuable information on Earth's internal structure. When you tap a crystalline wine glass with a spoon, the glass resonates at its natural frequency, which is in our audible range. If you fill the glass with wine and tap it again, the natural frequency of the sound wave it emits will have changed. One can obtain information on the structure and contents of the glass by mathematically analyzing the frequency at which the glass resonates. Similarly, information can be obtained for the internal structure of Earth after it has been set to ringing by a large earthquake. For example, mathematical analyses (using spherical harmonic analyses, discussed in chapter 1) of Earth's natural frequencies have provided the best estimate for the rigidity of Earth's inner core.

Two years after the great Chilean earthquake of 1960, astronomers noted that the surface of the Sun rose and fell with a period of nearly five minutes.[1] However, it was not until 1975 that these motions, which reflected solar resonances, were analyzed using harmonic analyses to provide crucial information on the Sun's internal structures and dynamics. The Sun, which contains 99.8 percent of the mass in our solar system,[2] was found to have three major divisions. About half the mass of the Sun is in its core, which has a radius of about a quarter of the Sun's total radius of 0.7 million km (about 0.4 million miles). Thermonuclear reactions, involving the fusion of hydrogen into helium (chap. 3), produce thermal energy in this core. The heat generated in these reactions is radiated upward through a radiation zone, the second major division, by the same process one is warmed by a campfire: heat is transferred by infrared light traveling as electromagnetic waves. The radiation continues upward until it encounters the convection zone, which begins at around 71 percent of the Sun's radius and ends at the Sun's surface. In this zone, heat is transferred by the convection of plasma, which, in this instance, is a gas primarily consisting of protons, electrons, and helium nuclei. (There are also a few heavier nuclei: the Sun's composition is about 70 percent hydrogen, 28 percent helium, and 2 percent heavier elements.) The Sun's surface exhibits a granular pattern, which is caused by turbulence in the convection zone. An individual convection cell, a region where plasma moves up in the center and descends at the edge, has an average diameter of 1,000 km (620 miles). A mental picture of a convection cell can be constructed by picturing a heater in the center of a small room. Air, warmed by the heater, rises to the ceiling, where it spreads out until it reaches the walls. There it descends to the floor to flow back toward the heater to complete the convection cell. The sizes and locations of convection cells on the Sun change rapidly, reflecting the turbulent nature of the Sun's convection zone.

*   *   *

The movement of ions (electrically charged atoms) and electrons in the convection zone produces electric currents, which generate a hierarchy of magnetic fields. These fields range in size from a global dipole field, which usually has an intensity somewhat larger than Earth's dipole field, to relatively small-sized fields less than that of an individual granule. Of these, the magnetic fields of sunspots are the most famous and infamous.

Sunspots are now known to be dark irregularly shaped areas on the solar surface with temperatures about 1,000°C cooler than the average temperature of the Sun's surface, which is near 6,000°C (about 11,000°F). Although reports of sunspots extend back more than two millennia, many of the earlier observations were subsequently discounted. Aristotle (384–322 BC) thought the heavens were perfect and unchanging. Because this would not be the case if the Sun had blemishes, there could be no sunspots. However, a large sunspot was reported in Europe near the time of Charlemagne's (Charles the Great's) death in 814 AD, long before the telescope was developed. Because everyone knew the Sun could not have blemishes, several years later this observation was attributed to the planet Mercury passing in front of the Sun. Science eventually can overturn erroneous beliefs, such as the Sun has no spots, but sometimes it requires an enormous amount of evidence to do so. In the case of sunspots, such evidence was not convincing until after Galileo developed the telescope and used it to see what first appeared to him to be clouds crossing the Sun's surface.[3]

The Italian scientist Galileo Galilei (1564–1642) had a mixed education. He entered the order of a monastery school as a novice against the wishes of his father. His father then moved him to Florence, where he again flirted with a career in religion before entering the University of Pisa to study mathematics. Eventually he left the university without a degree to return to Florence, where he studied mathematics outside a university setting. Although an unusual path, it led to a short-lived position in 1589 as a mathematics professor at the University of Pisa. Because his attacks on Aristotle made him unpopular with his colleagues, he was forced to leave in 1592. Nevertheless, he secured a professorship at another Italian university, the University of Padua.

Galileo developed the first telescope in 1609, which magnified objects nine times, a magnification similar to that found in modern-day birders' binoculars. He turned his telescope to the Moon and saw mountains rather than the smooth surface expected based on the teachings of Aristotle. He also discovered four moons orbiting Jupiter. (There are now known to be more than sixty moons orbiting Jupiter.) Jupiter and its moons probably appeared to Galileo as a miniature solar system. Such observations only made him more skeptical of the teachings of Aristotle.

Galileo was persecuted by the Roman Catholic Church, not for his observations of sunspots, but for his heliocentric (Sun at the center) view of the solar system. Scientists such as Copernicus and Kepler had supported a heliocentric picture before Galileo did, but geocentrism (Earth at the center of the solar system) was still widely accepted. Galileo was particularly vocal about it being wrong, contrary to the teachings of Aristotle and the Roman Catholic Church. In 1616 Cardinal Robert Bellarmine issued an edict forbidding Galileo from defending the heliocentric theory. Around 1632 Galileo published arguably his most famous book, *Dialogo sopra i due massimi sistemi del mondo, Tolemaico e Copernicano*, which involves a dialogue on geocentrism and heliocentrism. This book has a witty dialogue between three individuals, Salviati, Sagredo, and Simplicio. Salviati is a wise teacher who holds the same views as Galileo, Sagredo is an intelligent layman, and Simplicio, a supporter of the Aristotle view of the world, uses poor arguments remarkably similar to those used by the then-pope, Urban VIII. This was too much for Rome. An Inquisition quickly followed in 1633, which led to Galileo publicly recanting his support for a heliocentric solar system under the threat of excommunication and the death penalty. He remained under house arrest until he died in 1642. The Roman Catholic Church finds this an embarrassing episode, and Pope John Paul II officially apologized for it in 1992. In 2008 the Catholic Church erected a statute of Galileo inside the Vatican walls.

Arguments between geocentrists and heliocentrists persisted for more than a century after Nicolaus Copernicus advocated a heliocentric model in 1543. William Gilbert, in the sixth book of *De Magnete* (see chap. 1), argued that magnetic fields of heavenly bodies explained their motions, including Earth's daily rotation. Of course, this was all hypothesized before Newton developed his law of gravity. Although Gilbert was a heliocentrist, his views on magnetism were later used to support geocentrism. In the mid-seventeenth century, the Jesuit mathematician Athanasius Kircher argued against the idea that a spinning spherical magnet could force another in orbit around it. In 1645 the Jesuit scholar Jacques Grandamy showed that suspended spherical lodestones (magnetite spheres) made poor timepieces. These findings falsified Gilbert's claim that Earth's clock-like motions were due to magnetic effects. It appeared clear to such scholars that the underlying magnetic hypothesis put forth by Gilbert, a heliocentrist, was wrong, and thus, they reasoned, geocentrism must be correct.

I do not know why the geocentric view persisted so long, but I suspect it had to do with religion and other cultural influences. Perfection (for example, heaven) was supposed to lie above us in the stars, while imperfection (for example, hell) was suppose to lie below us in Earth's center. The notion of "up" and "down" seemed to depend on Earth being at the center of the universe. If the Sun were at the center, as proposed by some scientists, didn't that mean that "down" was toward the Sun? Such philosophical foundations needed to change before a heliocentric viewpoint could be widely accepted. Even the strongly religious Sir Isaac Newton, who was the first to establish a reasonable mathematical law for gravity, tried to determine the location of heaven and hell.[4]

We have accumulated considerable information on sunspots and their associated magnetic fields since the time of Galileo and Newton. For example, we now know that sunspots are relatively dark cold regions on the Sun; the central core of the sunspot (umbra) can be 2,000°C lower than regions outside the spot, while most of the sunspot (the penumbra, surrounding the umbra) has temperatures about 1,000°C less than outside. Although darker relative to other areas on the Sun, sunspots would appear blindingly bright if seen in isolation. Even though sunspots appear as small features on the Sun's surface (photosphere), the largest ones have diameters of over 100,000 km (62,000 miles); they can be more than ten times the size of Earth. They are regions of the most intense magnetic fields on the Sun, which can reach strengths of more than 1,000 times the intensity of the Sun's dipole magnetic field.[5]

Sunspots exhibit an 11-year cycle, which has been linked to an incredible number of diverse phenomena, including cycles in the gross national product, the stock market, animal populations, physical and mental health, animal behavior, climate, volcanism, and earthquakes. With the possible exception of climate (discussed in chap. 6), mainstream scientists view such linkages as pseudoscience. They typically indicate abuse of statistics or inappropriate application of Fourier series to analyze for periodicities (chap. 1). However, the 11-year periodicity, or quasi-periodicity, of sunspots has been well documented. It is the alleged periodicities of other phenomena (such as earthquakes) and their correlation with sunspot behavior that scientists abhor.

The first suggestion of periodicity in sunspot behavior can be traced to a German pharmacist and amateur astronomer, Heinrich Schwabe (1789–1875). Motivated by a short paper by Schwabe, the Swiss scientist Rudolf Wolf (1816–1893) compiled past observations of sunspots to find a quasi-periodic 11-year cycle; the periodicity varied mostly between 9 and 13 years, but averaged to 11 years. The beginning of a sunspot cycle occurs with the formation of sunspots in narrow latitudinal bands centered near 35° in both hemispheres of the Sun. Although sunspots come and go, the bands containing them drift toward the equator. The number of sunspots increases with this drift until the bands are between 15° and 20° latitude. Because the total number of sunspots in a year is usually greatest at this point in the cycle, scientists refer to this as the sunspot maximum. This is manifested in the Maunder butterfly diagram (upper graph in fig. 4.1), named after the husband and wife team of Edward and Annie Maunder, who published the first butterfly diagram in 1904. The lower graph in figure 4.1 shows how the total area occupied by sunspots, which is correlated with the average number of sunspots, varies from cycle to cycle. We will return to this figure when we discuss climate in chapter 6. The last solar maximum occurred during 2001 and 2002, and the next one (in cycle 24) is expected to occur around 2013. After solar maximum, the bands continue to drift toward the equator, the number of sunspots decreases, and eventually sunspots disappear near 5° latitude. The 11-year cycle renews itself when new sunspots form in bands with centers near 35° latitude. The pattern of sunspots forming at intermediate latitudes and drifting toward the equator is known as Spörer's law. It was discovered by Richard Carrington in 1861 and later refined by Gustav Spörer.

The temperature in a sunspot is substantially colder than its surroundings only to a depth of around 3,000 km (about 2,000 miles). However, the magnetic fields extend to a depth close to the bottom of the convection zone and provide some stability to the sunspot. Even with this stability, individual sunspots seldom last more than a week or two. You might wonder how sunspots can exhibit an 11-year cycle, while individual sunspots

DAILY SUNSPOT AREA AVERAGED OVER INDIVIDUAL SOLAR ROTATIONS

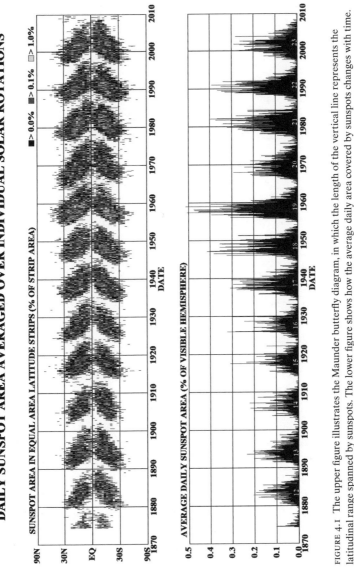

FIGURE 4.1 The upper figure illustrates the Maunder butterfly diagram, in which the length of the vertical line represents the latitudinal range spanned by sunspots. The lower figure shows how the average daily area covered by sunspots changes with time. Sunspot cycles in this figure run from cycle 11 in the middle of the eighteenth century to cycle 23. Cycle number 24 began in 2008. These figures illustrate the 11-year sunspot cycle periodicity, or quasi-periodicity. Figure modified from NASA images.

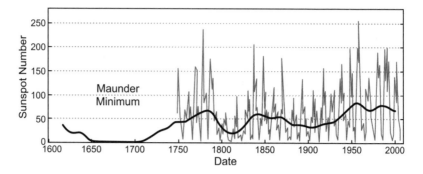

FIGURE 4.2 The number of sunspots varies from cycle to cycle. In a few cases, the maximum number of sunspots seen in a year exceeds 200. Few sunspots occurred between 1645 and 1715, the Maunder minimum. There are only sparse data prior to 1750. The solid black line represents an average of the sunspot activity and will be referred to in chapter 6. The original figure from *Wikimedia Commons* was redrawn by Beth Tully.

last such a short time. The sunspot cycle is somewhat similar to traffic on a freeway through a major city. Rush-hour traffic might only move at an average rate of 20 miles per hour or so for distances exceeding 30 miles. However, individual cars may have a short "lifetime" on the freeway because frustrated drivers exit the freeway after only traveling a short distance. Similarly, the lifetimes of individual sunspots are relatively short, even though they collectively travel vast distances on the Sun, taking 11 years or so to reach low latitudes.

Sunspots also exhibit a 22-year periodicity. Most sunspots occur in pairs, which are said to be "tilted" in that the lead sunspot (in the eastward direction of the Sun's rotation) is closer to the equator. The lead sunspot has one magnetic polarity, while the trailing sunspot has the opposite polarity. The magnetic field exits at one sunspot and enters the second sunspot in the pair. When the lead sunspot has "normal" polarity, the trailing sunspot has "reverse" polarity. The order of the polarity changes every 11 years at sunspot minimum: during the next sunspot cycle, the lead sunspot has reverse polarity and the trailing sunspot has normal polarity. That is, it requires 22 years on average, the Hale cycle, for the lead sunspot to return to its "original" polarity.[6]

This sunspot behavior is coupled to reversal of the Sun's dipole field, which occurs approximately every 11 years near sunspot maximum (not minimum, as is sometimes erroneously stated). Unlike Earth's dipole magnetic field, which reverses randomly in time, the solar dipole magnetic field reverses periodically about every 11 years. However, there may be times when this does not happen. Very few sunspots occurred during the interval extending from 1645 to 1715, the Maunder minimum, as illustrated in figure 4.2. The existence of this minimum is now well accepted, having been supported by isotope data and other phenomena, such as observations of auroras.[7] Although it is not known what the Sun's dipole did during the Maunder minimum, some scientists have speculated it maintained a constant polarity. We will return to the Maunder minimum in chapter 6 when climate is considered; sunspots are a time of high solar activity, and some geomagnetists have claimed the decrease in solar activity during the Maunder minimum was responsible for the Little Ice Age (chap. 6).

*   *   *

Sunspots are just one of many solar phenomena linked to an active dynamo. Unlike Earth, where the dynamo originates in the core, the solar dynamo originates in the outermost convection zone. Because of this, solar dynamo theory can more easily be tested by observations than Earth dynamo theory. For example, astronomers can observe magnetic field reversals while they happen—unlike for Earth, where their properties must be inferred from indirect measurements of rocks (chap. 2). Usually the Sun's global magnetic field is predominantly dipolar, but it becomes increasingly complex as it approaches solar maximum, and during reversal it becomes predominately nondipolar.

Astronomers can also determine the flow of the plasma in the Sun's convection zone far more precisely than the flow of the molten iron in Earth's core. Astronomers have long known that the Sun does not rotate at a uniform rate. The Jesuit priest Christopher Scheiner used the motion of sunspots at different solar latitudes to conclude in 1630 that the Sun rotates faster near the equator than at higher latitudes. Unlike a rigid body, a fluid body need not rotate at the same rate. The water at the sides of a canal, for example, flows slower than that at the center. A complete (eastward) rotation of the Sun's surface the (photosphere) near the equator requires about 25 Earth days. The photosphere's rotation gradually decreases with latitude to a minimum near the poles, where a complete revolution requires approximately 35 Earth days.

Although differential rotation at the Sun's surface has been long recognized, the internal differential rotation of the Sun was erroneously thought to occur on concentric cylinders, until helioseismology (seismology of the Sun) analyses were used. Helioseismic data now involve other types of solar oscillations than the up-and-down motions described earlier. The launching in 1995 of SOHO (the Solar and Heliospheric Observatory satellite operated by NASA and the European Space Agency) enabled scientists to obtain unprecedented measurements of the Sun, including its internal motions. Such data can even be used to detect sunspots on the opposite side of the Sun from Earth.

SOHO is one of many missions undertaken by NASA. In 2008 NASA listed 27 current missions, all of which are expensive and involve numerous scientists. Typically one scientist is in charge (called the principal investigator) of developing a scientific instrument and its use during a particular mission. Such scientists invest much time and energy into the development of an instrument, which may never actually be used. Compared to the high costs of missions, the fraction of funds devoted to the development of instruments for scientific research in the NASA budget is small. Therefore, NASA administrators fund many scientific projects for possible use on missions—many of which never occur. For example, Congress might ask NASA officials to scrap one mission in favor of another. Moreover, even when a mission does occur, things can go wrong—sometimes terribly so.

I do not know where the term "rocket scientist" came from, but I don't know anyone who claims to be one. There are space scientists, scientists who work on various aspects of the magnetic field and related problems in our solar system. They seem to be the closest to rocket scientists, because they sometimes use instruments in rockets and satellites to carry out their research. There are many engineers involved in the development of rockets. Are the engineers, who are technically involved in technology rather than science, the stereotypical "rocket scientists"? I don't know. In any case, scientists and engineers sometimes communicate poorly, and this can lead to disasters. One of the highly publicized failures of NASA came in September 1999 when the $125 million Mars Climate Orbiter crashed. The spacecraft was supposed to orbit Mars to relay information on its weather. However, confusion over units led to it crashing into the surface of Mars before transmitting any weather data. To my knowledge, there are only two countries besides the United States still extensively using English units for measurement: Liberia and Myanmar (Burma). The metric system is the choice of units used worldwide by scientists, and it is the choice of units used by NASA. By contrast, engineers in the United States typically use English units. The Lockheed Martin Corporation designed the critical navigational software for the Mars Climate Orbiter mission. In doing so, they used English units for force (pounds), which NASA scientists misinterpreted as metric units (Newtons). Although NASA and Lockheed Martin blamed each other following the crash of the Mars Climate Orbiter, it is clear that NASA needed more checks in their system to avoid such disasters. Apparently even rocket scientists can sometimes exhibit intellectual weaknesses.

Most of NASA missions have been remarkably successful. One of these missions involves the SOHO satellite, which is in a stationary orbit between the Sun and Earth. Because the gravitational attraction of Earth on SOHO is balanced by the gravitational attraction of the Sun, SOHO is essentially always in the same position relative to Earth and the Sun. (This balance also includes the centrifugal force.) The principal investigator of the Michelson Doppler Imager on SOHO is Phillip Scherrer, a bearded Stanford University professor who fits the stereotypical picture of a "rocket scientist." The helioseismology data from this instrument yield information on the solar resonances that have played a major role in the development of twenty-first-century dynamo models for the Sun.

Although solar dynamo theory is still sometimes described using the intuitive dynamo model developed by Eugene Parker (chap. 3), by 1980 theorists had become well aware that many of the solar magnetic field features differed from those of Earth, requiring the development of different dynamo models for Earth and the Sun. In particular, solar dynamo theory needed to explain the 11- and 22-year sunspot cycles and their clear connection to the Sun's global magnetic field, which reverses polarity during sunspot maximum a half cycle later than the polarity reversal of the lead (and trailing) sunspots. In 1987 Eugene Parker coined the phrase "the solar dynamo dilemma" to emphasize the then-inconsistencies between theory and observations: the input of helioseismology data for the flow of the Sun's electrically conducting plasma into dynamo models indicated that sunspots would travel in the opposite direction from that observed.

Modern dynamo models solve this problem by allowing much of the dynamo action (chap. 3) to occur in a thin zone, called the tachocline, at the bottom of the convection zone. Flow in this zone is in the opposite direction from the flow at the surface. Although not previously mentioned, dynamo theory has long incorporated the fact that magnetic fields exert a pressure on their surroundings. (This pressure is proportional to the square of the magnetic field.) This is used in twenty-first-century dynamo models to produce most properties of the sunspot cycle. Just as air in a hot-air balloon pushes outward, causing the balloon to become buoyant, magnetic fields that are bunched together in the tachocline produce buoyant regions. Some of these regions rise to the surface, where they are manifested as sunspots. They rise quickly enough to retain their equatorial direction of movement, which is inherited near the bottom of the convection zone. Scherrer and others now advocate that sunspots originate at a depth near 200,000 km (124,000 miles). As they rise, they cool in a manner similar to how air cools as it flows up a mountain slope. Sunspots only become substantially cooler than their surroundings when they are within 3,000 km or so of the Sun's surface.

Present-day solar dynamo theory is similar in its broad aspects, but not in the details, to Earth's dynamo theory.[8] In both cases, an initial field of unknown origin is magnified to produce the magnetic fields observed on the Sun and Earth (chap. 3). In the twenty-first century, many so-called self-consistent three-dimensional dynamo models exist for the Sun. They usually involve differential rotation in the tachocline to produce strong horizontal (toroidal) magnetic fields in a manner similar to that shown in figure 3.2. Regions of upwelling in the convection zone result in the formation of closed magnetic field loops, which merge to produce a dipole magnetic field and the associated sunspot magnetic fields.

Although there are many dynamo models for the Sun, none is adequate. As for Earth, no self-consistent dynamo model has been developed that adequately treats turbulence (chap. 3). In particular, there is no adequate explanation for the complex relationship between the sunspot cycles and the Sun's global dipole field. Nevertheless, as for Earth, scientists have no doubt that dynamo action is responsible for producing the Sun's main magnetic field.

*   *   *

Heat is transferred from the bottom of the convection zone, where it is at a relatively high temperature, to the Sun's surface, which is colder. It is this transfer of heat by convection that drives the solar dynamo. While the details of this heat transfer are poorly known and will remain so until the dynamo process is better understood, even the basic mechanism of heat transfer is not known in the solar corona.

The glow seen surrounding the Sun during a total solar eclipse is called the corona ("crown," from Latin). The corona is sometimes referred to as the "Sun's atmosphere." However, unlike Earth's atmosphere, its density of ions is typically a trillion times smaller than in the photosphere. The corona, which has no upper boundary, is not your stereotypical "atmosphere."

One of the Sun's great mysteries is that its corona can be hundreds of times hotter than the 6,000°C temperature of its surface. Although the temperatures vary in time and with location in the corona, they typically exceed a million degrees. The famous second law of thermodynamics requires heat, a form of energy, to flow from hot regions to cold ones. How then, does the colder Sun's surface transfer heat to the hotter corona? When you raise the temperature in your house, the molecules of air travel faster. When they impact your body, they convert their energy of motion (kinetic energy) into heat. The heat in the corona is proportional to the speed (or more precisely the square of the speed) of its particles (ions and electrons). The temperature mystery can be restated: how can the speed of particles in the corona be so much greater than in the photosphere? So far scientists have only been able to produce speculative models.

As an aside, even though the particles in the corona are traveling rapidly enough to produce enormous temperatures, you would freeze if you were to find yourself in a room with similar hot plasma. There would be far too few particles impacting your body to warm you.

The first thing a theorist should do before starting calculations is to make sure the phenomenon she is trying to explain is real. In the case of the Sun's temperature mystery, it is prudent to make sure that the temperature, or equivalently the velocity, measurements are correct. When ions travel at high speeds, they emit X-rays. In 1991 the Japanese Institute of Space and Astronautical Science launched the satellite Yohkoh ("sunbeam") from southern Japan to measure X-rays coming from the Sun's surface and corona. Results obtained from Yohkoh, which ceased operations at the end of 2001, confirmed the high temperatures in the corona. The temperature mystery is a real phenomenon.

What's next? How do we construct a theory that does not violate the laws of thermodynamics, which require that energy be conserved? A key is provided by the recognition that no fundamental law is violated if some other form of energy is converted to heat in the corona. Most scientists working on this problem now believe electromagnetic energy is being converted to heat. I suspect you have experienced the transformation of electromagnetic energy into heat. A short circuit in electrical wiring produces heat—sometimes enough to cause a fire. But, to quote a cliché, the devil is in the details.

One promising mechanism to convert magnetic energy into heat involves magnetic reconnection, a phenomenon that will also be useful to us when we try to understand the origin of magnetic storms. It is common to use magnetic field lines, which are not real quantities, to provide physical insight into reconnection and related phenomena. Magnetic field lines

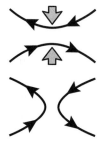

FIGURE 4.3 The upper figure shows magnetic field lines (in black) before merging. Merging occurs in the direction indicated by the gray arrows. The lower figure shows the field lines after merging and reconnection have occurred. Only parts of the field lines representing the magnetic field loops are shown. Figure drawn by Beth Tully.

give the direction of a magnetic field. They form closed loops around an electric current (see fig. 1.2). These loops will change their sizes and shapes to reflect the evolution of a magnetic field. For example, two small loops can merge to form a larger one. In other words, smaller-size magnetic fields can combine to produce larger ones. Sometimes the merging of loops is reconfigured in a complex way. In this case, original loops might break into pieces and recombine to form a very different group of loops. Magnetic reconnection is common during such a reconfiguration. Magnetic reconnection refers to the breaking apart of two magnetic loops and the formation of two new ones. Figure 4.3 shows the merging of two magnetic field loops, reconnection, and the formation of new magnetic loops. Only parts of the original loops are shown in both the upper and lower figures of 4.3.

Magnetic field lines can be considered to have a tension and can be pictured as rubber bands (chap. 3). The flow of relatively dense conducting fluid (plasma) can push (gray arrows in fig. 4.3) the field lines together until they merge. The magnetic field lines will form an X at the instant of reconnection. After reconnection occurs, the field lines (rubber bands) can reduce their tension by straightening. In this process, they accelerate plasma (to the right and left for the case shown in the lower illustration of fig. 4.3). The reorganizing of field lines, including reconnection, close to the Sun's surface is speculated to fling particles in plasma into the corona much like a slingshot would. Several other mechanisms leading to high velocities of particles in the corona have also been proposed, and it seems likely that multiple mechanisms are at work.[9]

* * *

In 1958 Eugene Parker predicted that the ions and electrons in the corona were hot enough to overcome gravity and stream away in all directions from the Sun in a "solar wind." Ninety-five percent of the ions in the solar wind are protons, 4 percent helium nuclei, and 1 percent nuclei of heavier elements. Because ions and electrons are good conductors of electricity, Parker used magnetohydrodynamic theory (chap. 3) to suggest that the solar wind also carried with it the Sun's magnetic field. This solar wind, subsequently confirmed by numerous measurements, ultimately reaches speeds of 200 to 1,000 km per second (124 to 620 miles per second)—typically more than 1,000 times the speed of sound in air on Earth. On average, when the solar wind approaches Earth, it contains about 5 ions per cubic centimeter and a magnetic field, referred to as the interplanetary magnetic field (IMF), with an intensity around 1/1,000 of the magnetic field at Earth's surface. Space is not the vacuum most scientists had assumed during the first half of the twentieth century: it contains ions, electrons, and electromagnetic fields.

Just as wind in our atmosphere can erode and alter Earth's landscape, the solar wind can erode planetary atmospheres. It exerts a pressure (from the ions and the IMF) on any object it encounters in its path. The solar wind flows around Earth, and its internally produced magnetic field flows much like water does around a rock in a stream. The locations where the solar wind and Earth's magnetic field are in balance define the boundary of Earth's so-called magnetosphere. That is, at the boundary of the magnetosphere, the magnetopause, Earth's magnetic field pushes back on the solar wind with an equal and opposite pressure. The magnetopause is about 10 Earth radii, about 63,710 km (40,000 miles), from Earth's center in the direction of the Sun. (Space scientists often speak of distances in terms of Earth radii. One Earth radius is equal to 6,371 km.) The solar wind also affects the geometry of the magnetosphere's long drawn-out tail, which occurs on the night-side of Earth (fig. 4.4).

The magnetosphere extends inward to the top of the ionosphere, approximately 500 km above Earth's surface. Inside lies our atmosphere and our planet. The magnetosphere's shape is nothing like a sphere. Envisage a rock (Earth) in a stream (the solar wind). The water diverts around the rock on the upstream side, and it forms a long tail on the downstream side. Although the tail is long, it is not clear where it ends. Because the water in the stream fluctuates, so does the flow around the rock.

Because both the solar wind and Earth's magnetic field change over time, the distance to the magnetopause is variable. In particular, the density and velocity of the solar wind changes rapidly, affecting the magnetopause's position. For example, it can be compressed to less than 6 Earth radii

FIGURE 4.4 An artist's view of the solar wind and Earth's magnetosphere. This wind—which consists primarily of protons, electrons, and magnetic field—travels in all directions away from the Sun. The magnetosphere boundary occurs where the pressure of the solar wind is balanced by the pressure of Earth's internal magnetic field. Figure modified from NASA images.

during a magnetic storm, a time when the solar wind is particularly intense, as will be elaborated on momentarily. It can also expand—greatly in rare instances. On May 11, 1999, the solar wind density dropped to 2 percent of normal. The pressure on the magnetosphere dropped accordingly, and the magnetosphere ballooned out to nearly 380,000 km (235,000 miles). The magnetic field at Earth's surface became very quiet. Magnetic storms do not occur during such times.

Earth is not unique in having a magnetosphere. All planets with dynamos (chap. 3) exhibit magnetospheres. Even planets without active dynamos but with intense magnetization, such as Mars, are sometimes said to possess "mini magnetospheres," which help protect the most strongly magnetized sections of their surfaces from the solar wind. Although Venus lacks a magnetic field, it has an ionosphere that protects the inner atmosphere from the solar wind. Because solar magnetic fields can induce electric currents in Venus's ionosphere, some scientists refer to Venus as having an induced magnetosphere.

The solar wind is also used to define the boundary of our solar system, which is about three times the distance from the Sun to the dwarf planet Pluto. The solar system boundary used to be defined as only extending just past Pluto, which is on average about 5.9 billion km (about 3.7 billion

miles) from the Sun. The new definition of the boundary, called the heliopause, is where the solar wind cannot be distinguished from the particles
and magnetic fields of interstellar space. *Voyagers 1* and *2* are expected to
cross the heliopause within the next decade. These spacecraft, which were
launched in 1977 to investigate the outer solar system, are still transmit-
ting data back to NASA.

Our galaxy, the Milky Way, also appears to have a wind that carries
with it a magnetic field (chap. 3). When this wind is directed toward our
solar system, the heliopause is pushed inward toward the Sun. Measurements made on *Voyager 1*, which traveled southward, and *Voyager 2*, which
traveled northward, indicate that the distance to the heliopause is about
1.6 billion km closer to the south than the north.[10] That is, the heliopause
appears to be deformed by a galactic wind flowing northward.

Let's examine the effects of the solar wind on Earth. Earth's atmosphere
resides within, and is protected by, our magnetosphere. If it were not for
the magnetosphere, our atmosphere would be substantially reduced because of sputtering, the erosion and chemical alteration of our atmosphere
that would occur when particles in the solar wind bombard atmospheric
atoms and send many of them into space. Without a protective magnetic
field, our atmosphere would look more like Mars's rarefied atmosphere.
Without it there would be no life on Earth, with the possible exception of
some microbes.

The composition of Earth's dry atmosphere (when no water is present)
consists of 78 percent nitrogen and 21 percent oxygen and small amounts
of other gases. Although the air becomes less dense with elevation, the
composition of the atmosphere varies little to a height of 100 km (62 miles).
Less than 1/10,000 (by mass) of our atmosphere lies above 100 km, while
more than 90 percent lies below 12 km or so—the upper boundary of the
troposphere in which our weather occurs.[11]

The ionosphere constitutes the upper part of the atmosphere, and it plays
an important role in radio transmission. The lower ionosphere, which begins near a height of 50 km, extends upward a few hundred kilometers. An
ion is produced in the ionosphere when ultraviolet radiation from the Sun
knocks an electron out of a neutral gas atom (one with no electric charge)
during a collision. The ion can later recombine with another electron to
become an electrically neutral atom again. Few ions are present in the
lower part of the ionosphere where recombination is rapid. Few ions occur
in the upper reaches of the ionosphere for a different reason: the number
of atoms available for ionization is relatively small. A broad maximum

in ionization is present near 300 km, where the rate of ionization by ultra-violet light greatly exceeds the rate of recombination.

In 1899 Nikola Tesla carried out experiments to see if the ionosphere could be used to transmit low-frequency electromagnetic signals without using wires. The magnetic field in mks (meter-kilograms-second) units is measured in tesla in his honor. It was once thought that radio waves could not be transmitted great distances because they would follow a straight line and the surface of Earth was curved. However, in 1901 Guglielmo Marconi received a radio signal in Newfoundland sent from England. He used an antenna attached to a kite for reception. Because the radio signal reflected off the underside of the ionosphere, it was transmitted over a much larger distance than many people thought was possible at that time.

Two flattened doughnut-shaped regions, the Van Allen belts, lie above the ionosphere, as shown in figure 4.5. They circle Earth and die out at high latitudes. Ions and electrons are more abundant in these belts than in their surroundings. Although there were speculations made during the first half of the twentieth century regarding the existence of radiation belts, they were not confirmed until the first NASA Explorer missions (*Explorers 1* and *3*) began in 1958 under the direction of James Van Allen (1914–2006). The inner Van Allen belt, lying between 700 and 10,000 km altitude, is dominated by high-energy protons (ones traveling with high velocities).[12] The outer belt begins about 20,000 km, has its greatest intensity at about 30,000 km, and tapers off above this maximum. This belt primarily consists of electrons, although it also contains some protons and heavier ions.

The inner belt is the primary contributor of protons to a "ring current," an electric current that flows around the Earth's rotation axis. The Van Allen belts and the ring current play important roles in magnetic storms and substorms. NASA launched 73 Explorer spacecraft in the previous century to explore these belts and related phenomena. This exploration continues in this century, most recently with the launch of five THEMIS (Time History of Events and Macroscale Interactions during Substorms) satellites in 2007.

However before we explore the ring current and magnetic storms, I wish to mention some common misconceptions associated with the Van Allen radiation belts.

The word "radiation" has a bad connotation. Some of this is justified, as evidenced by the horrible aftereffects of the atomic bombing of Hiroshima and Nagasaki and the reactor explosion at Chernobyl in Ukraine. However, it is essentially impossible to avoid all radiation, since many

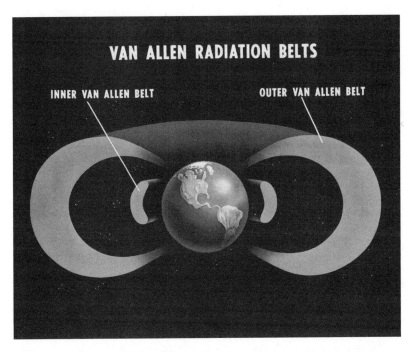

FIGURE 4.5 Artist's view of the Van Allen radiation belts. Figure from NASA images.

rocks contain radioactive minerals and Earth is constantly bombarded by high-energy particles from space called cosmic rays (discussed more in chap. 6). In the case of the Van Allen radiation belts, the most dangerous particles are high-energy protons, which can produce significant damage to instruments and humans during prolonged exposure. Spacecraft and astronauts try to minimize time spent in the Van Allen belts, particularly the lower one, to minimize such damage. To assess the effects of radiation, every Apollo astronaut wore a radiation dosimeter. Astronauts on *Apollo 14* (1971) received the greatest exposure to radiation, but still less than a percent of a fatal dose. Contrary to the claims of some conspiracy theories, there is no evidence that NASA tried to cover up the dangers of the radiation in these belts. Some conspiracy theories even claim it was fatal to pass through these belts. Thus, it is concluded, no astronaut did and the lunar landings were faked. Although fraud occurs in science and about 6 percent of Americans still believe NASA staged the lunar landings (according to a Gallup Poll taken in 1999 to coincide with the thirtieth anniversary of *Apollo's 11*'s landing; an additional 5 percent were unsure),

I would be surprised if anyone reading this book accepts such a conspiracy theory.

* * *

Alexander von Humboldt (1769–1859) was a naturalist-adventurer, well known in his time for explorations of the rain forests in Venezuela. He also strongly promoted science during his many travels. He convinced the governments of many countries, including those in the British Commonwealth (ranging from Canada to Australia) and some outside the Commonwealth, such as Russia, to establish magnetic field observatories. Because some magnetic disturbances were recorded (on magnetograms) at these observatories at nearly the same time, he thought the disturbances might originate in outer space. He coined the term "magnetic storm" to describe them.

Magnetic storms are often called geomagnetic storms to indicate that they are magnetic storms on Earth and not some other planet. For example, Jupiter and Saturn have large magnetic storms; they also have large radiation belts.

On an October morning in 2005, I took in the spectacular view of the Alaskan range, including Mount Denali (McKinley), from the University of Alaska campus. I was visiting the university to deliver the Chapman Lectures, named after Sydney Chapman (1888–1970), a scientific giant who had coined the name "geomagnetism." He had published over four hundred research papers and published two encyclopedia-like volumes entitled *Geomagnetism* (coauthored with Julius Bartels in 1940). In 1930 he and one of his students, Vincent Ferraro, first proposed the existence of a giant cavity containing Earth's magnetic field, its atmosphere, and our planet. This became known as the Chapman-Ferraro cavity, which resembles the entity we now call the magnetosphere. Chapman also believed that the Sun occasionally emitted plasma that impacted Earth's (now-called) magnetosphere to create magnetic storms.

We now recognize magnetic storms are worldwide events that last from hours to days. These storms are described as having a "sudden commencement," a phrase first used by Chapman, because they start with a drop in magnetic intensity following the arrival of a fast-moving cloud of plasma from the Sun. Space scientists now use magnetic indices, such as the Dst index ("D" for disturbance and "st" for storm), to identify magnetic storms.[13] A typical magnetic storm involves a drop of about 1/100 to

1/1,000 in the intensity of the magnetic field measured at Earth's surface. This decrease occurs over 6 to 24 hours, followed by a gradual recovery over 1 to 4 days.

In 1746 a French scientist, Jean-Antoine Nollet, was interested in how fast electricity could be transmitted. Nollet, a monk, persuaded nearly two hundred other monks to form a circle nearly a kilometer in circumference. The monks were connected by pieces of iron. This metal-and-human chain was connected to a primitive battery (consisting of Leyden jars). When the battery was discharged, every monk was shocked at almost the same time. Clearly electricity traveled very fast. This experiment helped pave the way for others to explore ways electric currents might be useful for communication.

Years later, in 1832, the first telegraph was developed in Europe, followed by the development of one in North America in 1836. In September 1859, a massive eruption occurred on the Sun called a corona mass ejection. It carried billions of tons of plasma and embedded magnetic field. The perfect magnetic storm arrived at Earth in less than a day. It was so strong that the Van Allen belts were temporarily eliminated. Michael Faraday had already shown that changes in magnetic fields induce electric currents in conductors and semiconductors. The 1859 superstorm caused electric currents to be induced in the telegraph wires. These currents traveled through the ground to complete the electric circuit. The currents were sufficient to short out the newly developed telegraph lines in Europe and North America, causing numerous fires. Auroras were documented as far south as Rome, Cuba, and even Hawaii.

A large drop in a magnetic field indicates the onset of an intense magnetic storm. (A large drop means one greater than 250 nanotesla or so in the Dst index; that is, a decrease greater than about 5 percent of Earth's magnetic field.) Intense storms have caused major malfunctions of communication satellites, disrupted radio and television, affected GPS service, and caused major power outages. For example, storms produce radio blackouts because of increased ionization in the ionosphere, which leads to strong absorption of radio waves on the sunlit side. The Halloween magnetic storm of 2003 damaged more than one-third of NASA's satellites. Currently more than two hundred communication satellites are operating. The scientists Sten Odenwald and James Green used computer analyses to conclude in a 2008 *Scientific American* article that if a storm similar to the one of 1859 happened today, it would produce damage exceeding $20 billion. The potential financial cost caused by magnetic storms is large and growing.

One of the worries NASA has in returning to the Moon to establish a lunar base is the danger of solar eruptions. For example, an X-class solar flare, a designation given to the most violent flares, occurred in 2005 that accelerated protons to extremely high speeds and produced a days-long proton storm on the Moon. The Moon essentially has no magnetic field to deflect, or atmosphere to absorb, such particles. In contrast, our magnetosphere and especially our atmosphere offer us substantial protection from such storms. Approximately a dozen centimeters (5 inches) of water would have been required to absorb most of the high-energy protons of the 2005 storm. A space suit containing enough material to protect an astronaut would be too heavy to wear—even on the Moon. Any astronaut walking in a modern-day thin-skinned space suit at the time of a magnetic storm like the one in 2005 would likely have received a fatal dose of radiation. If a lunar base is established, it will have to provide sufficient shielding to protect astronauts. Because of the extra mass of material required to protect astronauts during a space flight, it is extremely difficult, and some scientists argue impossible, to offer astronauts much of a guarantee of safety.[14] In 2006 Eugene Parker published an article in *Scientific American* indicating that cosmic radiation alone can produce serious tissue damage and possibly even death to any astronaut on a Mars voyage. Parker argues that no known technology exists today or in the foreseeable future that can reliably protect astronauts on a Mars voyage.

Although there is no consensus on the details of what produces magnetic storms, a broad picture of storm origin has emerged. The Sun is more active near sunspot maximum, a time when flares and coronal mass ejections (CMEs) are observed to most frequently erupt. When flares or CMEs explode, they heat plasma to millions of degrees and accelerate protons (and a few heavier ions) and electrons to very high velocities. These particles carry along the local magnetic field, the one present at the site of the eruption. For example, a CME, which is typically larger than a flare, can contribute a few billion tons of particles and embedded magnetic fields to the solar wind. A large flare or CME produces a shock wave in the solar wind, which can travel so quickly that it impacts Earth's magnetosphere, a distance of 150 million km (93 million miles), only one or two days later. This shock wave rapidly compresses Earth's magnetosphere, which sometimes leads to a magnetic storm.

However, a rapid compression of the magnetosphere is insufficient by itself to create a magnetic storm. In 1961 James Dungey in England, the 1991 Fleming medalist, recognized that a storm also requires magnetic reconnection, which allows protons and electrons the easiest entry into our magnetosphere (fig. 4.3). In this instance, reconnection refers to the connection of Earth's magnetic field lines with those in the interplanetary magnetic field. Without reconnection, most ions and electrons are deflected near the boundary of the magnetosphere; they do not contribute to the currents within our magnetosphere.

Figure 4.3 shows why reconnection is easiest when the polarity of Earth's magnetic field is opposite to the interplanetary magnetic field. One can visualize the upper field line in the top figure as coming from the Sun and the lower one from the magnetosphere. Because Earth's present-day magnetic field has a northward component, reconnection occurs when the magnetic field carried by the solar wind has a southward component. If the two fields were pointing in the same direction, they would repel each other.

As elaborated on shortly, when reconnection occurs, many of the electrons and ions entering our magnetosphere end up spiraling along magnetic field lines. According to the Maxwell equations, charged particles experience a force perpendicular to both their direction of motion and the magnetic field direction. This produces a relatively slow "drift" of electrons and ions perpendicular to the magnetic field lines.[15] Positively charged ions travel westward and electrons travel eastward to produce a westward-traveling "ring current" in the magnetosphere at an altitude between about 20,000 and 30,000 km.

The magnetic field associated with this ring current is opposite to the dipole field produced in Earth's core (chap. 3). Therefore, when the ring current suddenly increases, a decrease in the geomagnetic field occurs, to produce a magnetic storm. The ring current subsequently returns to normal over one to several days, the recovery phase of the storm. Although measurements have confirmed this broad picture of magnetic storms, the details of the growth and decay of the ring current are considered an outstanding unsolved problem in magnetospheric physics.

On a September Monday in 2003, a massive CME erupted on the surface of the Sun and was recorded about a day later at the satellite SOHO (Solar and Heliospheric Observatory). A warning was issued for the possibility of a giant magnetic storm hitting Earth on Wednesday. Although this was one of the most violent solar eruptions ever recorded, magnetic reconnection did not occur and earthlings suffered no adverse effects. A few months later, the Halloween storm hit, reconnection occurred, and many satellites sustained substantial damage. Scientists did not have the technology (and still don't) to predict precisely when magnetic reconnection will occur: they are unable to determine accurately enough the polarity of the magnetic field at the site of the eruption of the flare or CME and its subsequent evolution.

Although space scientists cannot accurately predict all magnetic storms, the use of satellites enables them to predict them reasonably well about half the time. The other half of the time predictions fail, primarily because magnetic reconnection does not occur. Still an advance warning, such as can be provided by SOHO, can allow technicians to position satellites to minimize adverse effects, let astronauts try to protect themselves, and allow administrators to take steps to minimize power outages.

* * *

In addition to magnetic storms, space scientists refer to substorms. Chapman thought that substorms, which typically last only a few hours, were components of magnetic storms. While this is often the case, one of Chapman's students, Syun-Ichi Akasofu, subsequently showed that substorms could also occur independently of magnetic storms.

At the time of my visit, Akasofu, who received the Fleming Medal in 1979, was the director of the International Arctic Research Center at the University of Alaska in Fairbanks. After he retired as director in 2007, the building housing my temporary office was renamed the Akasofu building. On Wednesdays I met with Akasofu to discuss geomagnetism and afterward to have lunch. Akasofu pointed out that the University of Alaska issues daily forecasts that predict the intensity and location of auroras. An aurora is the best-recognized phenomenon of substorms.

It is easy to understand why auroras are often described as one of the world's seven natural wonders. They exhibit a wide range of colors and occur in a variety of forms, most commonly curtain and ribbon structures. Waves sweep the brilliantly colored curtains and ribbons, which occasionally fade away only to appear elsewhere a short time later. My wife and I once watched one such display of northern lights (aurora borealis; in the Southern Hemisphere, they are called the aurora australis) in a small game reserve behind the apartment complex we were housed in during our stay in Fairbanks. I looked down from this mesmerizing show by nature to notice a few horses were eating nearby. I wondered what horses were doing on a game reserve? When I turned on my flashlight to investigate further, the horses morphed into moose. We retreated to the safety of our apartment. Watching auroras in Alaska is not without its hazards.

Akasofu told me of many myths surrounding auroras. An old Scandinavian name for aurora translates to "herring flash," because the northern lights were thought to be reflections in the sky of large schools of herring traveling in the polar waters. The Algonquin aborigines of eastern Canada identified auroras as the lights of their ancestors dancing around a ceremonial fire. Some prospectors during the Klondike gold rush thought auroras were the reflections of a gigantic mother lode of gold.

Scientists often refer to auroras as polar auroras: they tend to form an oval at a distance of 2,000 to 3,000 km from the magnetic poles, for reasons to be given momentarily. Shortly after the beginning of the twentieth century, the Norwegian Kristian Birkeland (1867–1917) correctly suggested that auroras were caused by fast-moving electrons colliding with atoms and molecules in the high atmosphere. He carried out experiments that sent a beam of electrons toward a magnetized sphere. In 1908 he published the results: electrons tend to follow the magnetic field lines. Field-aligned currents (which travel in the opposite direction from electrons, because currents were defined to be carried by positive charges) have been confirmed and are now called Birkeland currents in his honor. Birkeland's contributions have not gone unnoticed in Norway: his picture appears on the front of the 200-kroner bank note.

Shortly after Birkeland published his experiments, the great mathematician Henri Poincaré (1854–1912) showed that electrons would spiral along magnetic field lines. Scientists now know electrons typically circle field lines thousands of times every second. Poincaré also calculated that the electrons would be reflected near the magnetic poles where magnetic field lines converge.[16] Modern measurements indicate that electrons travel from the north polar region to the southern one and back in about 1/10 of a second. Electrons continue to be trapped on magnetic field lines, bouncing between polar latitudes in both hemispheres, until they collide with atoms in the ionosphere to produce aurora light.

Auroras are manifestations of substorms, which are not well understood. Akasofu told me that the magnetosphere acts like a dynamo in which magnetic fields are amplified and electrons are accelerated. Just how this works is not clear. Because of such uncertainties, it is often convenient to refer to the movement of energy, rather than the specification of the movements of charged particles and magnetic fields. For example, there is widespread agreement that magnetic substorms involve the temporary storage of energy in the tail end of the magnetosphere, the magnetotail. Following storage in the magnetotail, energy is released inward to produce a substorm. However, the details of the processes involved are not known, and speculative models are hotly debated.

One leading model postulates that energy is stored in the magnetotail in a plasmoid, a large (bigger than Earth) flattened ballooned-shaped region in which ions and electrons are confined by a magnetic field. When the pressure on this plasmoid becomes sufficiently strong during a substorm, it is swept away from Earth by the solar wind. This involves magnetic reconfiguration, including reconnection. The magnetic field on the side of the plasmoid closest to Earth becomes unattached to the plasmoid. This magnetic field then moves toward Earth, flinging charged particles ahead of it, just as a slingshot (magnetic field line) would fling a rock (particle) when released. Many of these particles eventually end up spiraling along Earth's magnetic field lines.

The five THEMIS satellites launched in 2007 are currently carrying out measurements to test this theoretical picture. In any case, it has been shown that as the electrons near the geomagnetic poles, the magnetic field strength becomes strong enough to "reflect" them. The electrons then travel in the opposite direction along the field line until strong magnetic fields reflect them near the opposite magnetic pole. They bounce back and forth between hemispheres along magnetic field lines until they collide with atoms and molecules in the upper atmosphere to produce auroras.

Substorms are sometimes described using an analogy with an unstably balanced bucket hanging from a wire. Water (energy) flows into the bucket during a magnetic storm until the bucket turns over, spilling its contents in a substorm. The bucket then rights itself until incoming water (energy) again makes it unstable. This analogy is supported by observations confirming that the sudden flow of ions and electrons from the magnetotail is correlated with a sudden increase in aurora activity. Let's modify this analogy by placing a small hole in the bucket from which water (energy) drains. The drainage of water through the small hole is insufficient to stop the bucket from becoming unstable during a magnetic storm when it receives a rapid input of energy (water). However, the near continual loss of energy through the bottom of the bucket at other times still can lead to aurora activity, even though there may not always be enough light produced to see with the naked eye because the flow of energy from the magnetotail (where the bucket in this analogy resides) is small.

Many electrons released from the magnetosphere's tail during a substorm end up spiraling along Earth's magnetic field lines. Electrons travel to their lowest altitudes in the polar regions because the magnetic field lines become steeper as they approach the magnetic poles (see fig. 1.1). Since the magnetic field intensity also increases as the electrons approach the poles, the electrons are typically reflected back along the field lines on which they entered the upper atmosphere. Higher-energy electrons (with higher velocities) penetrate deeper into the atmosphere than lower-energy electrons before they are reflected. On occasion, some electrons are not reflected but collide with atoms and molecules at altitudes between 100 km (62 miles) and 300 km (186 miles). This increases the energy stored in atoms and molecules involved in the collisions, which is sometimes re-emitted a fraction of a second later in the form of aurora light.[17] Oxygen emits a white-green light, common of auroras, about 0.7 of a second after an electron has collided with it. Nitrogen usually emits ultraviolet or red light, depending on the details of the interaction, a nanosecond (a billionth of a second) after a collision with an electron. Because auroras form at nearly a constant magnetic field inclination, they produce an oval about the magnetic north (and south) pole. Larger substorms produce more collisions than smaller ones and more intense auroras.

The colors produced in auroras depend on such details as the amount of energy an electron has (which is proportional to the square of the electron's velocity) and the atom or molecule the electron hits. Akasofu often describes auroras by making an analogy to an oscilloscope, such as you may have seen during an electrocardiogram recording. The heartbeat is converted to an electron beam, which is directed to the oscilloscope's screen, which is coated with florescent material. This material emits light when impacted by electrons. One can change the colors by changing the type of florescent material. Aurora ovals are larger during major substorms and are smaller for lesser ones. During huge storms, aurora ovals expand so much they can be seen in temperate, and on much rarer occasions in tropical, latitudes.

Auroras are now known to occur in ovals about the magnetic poles of all planets with strong magnetic fields. Even Venus, which has no magnetic field, exhibits something akin to auroras: defused light has been observed during magnetic storms associated with electrically charged particles colliding with atoms in Venus's atmosphere.

In 1962 the American government conducted a high-altitude nuclear test, Starfish Prime, before a ban on such testing was put in place. Our Atomic Energy Commission detonated a nuclear warhead at an altitude of 400 km above the tropical Johnson Atoll in the Pacific Ocean. This caused malfunctions of radio and TV, knocked out streetlights, and fused power

lines in Hawaii, a distance of 1,400 km away. Several satellites were damaged or destroyed, and one could observe auroras in the tropical Pacific
for a few minutes. Not all substorms begin with solar eruptions.

*   *   *

The strength of magnetic storms and substorms should change when
Earth's internal magnetic field weakens, as it has been doing for the past
2,000 years. When the pressure from Earth's dynamo-produced field is
lower, the solar wind will push the boundary of the magnetosphere closer
to Earth. Small changes in intensity can have large effects because the
pressure associated with magnetic fields depends on the square of the field
strength. When the magnetic field intensity is cut in half, the pressure is
reduced to one-quarter. During a polarity reversal (chap. 2), the intensity
would decrease to about a quarter of its average value, presumably greatly
enhancing the effects of storms and substorms. Nevertheless, surprisingly
little detailed research has been done on these subjects.

On occasion, I have taught dynamo theory to small classes of space science graduate students and faculty at the University of Washington. The
first time I did this I was surprised when a faculty member interrupted to
say I was not using MHD (magnetohydrodynamic) theory. Upon further
discussion, he modified this to say I was not using "ideal MHD" theory. I
had no idea what he was talking about. Eventually he explained that I was
allowing the diffusion of magnetic fields to occur; in ideal MHD theory,
the diffusion of fields is not permitted. I explained we refer to this as the
"frozen in flux" approximation in dynamo theory (chap. 3). Solid-earth
magnetists and space physicists sometimes call the same mathematically
derived theories by different names. The theories are sometimes independently derived and published in different scientific journals. It is uncommon
for geomagnetists working on the magnetosphere to discuss their science
with geomagnetists working on Earth magnetism. It is so rare that organizations, such as the American Geophysical Union, do not even worry about
scheduling scientific sessions dealing with topics such as magnetic storms at
the same time sessions on dynamo theory are occurring. While these two
subfields of magnetism appear very similar to most people, including many
scientists, the degree of specialization and jargon makes it extremely difficult at times for a specialist in one discipline to understand a scientific talk
or read a scientific paper in the other discipline. Consequently important
problems at the interface of the two disciplines often are neglected.

It may surprise you, but the shape of the magnetosphere and its internal magnetic configuration have never been realistically modeled for a magnetic field reversal (chap. 2). Nor has the magnetosphere been adequately modeled for a lower-intensity magnetic field, such as occurred about 6,000 years ago. The absence of such models has consequences even for scientists not working in magnetism. For example, geochemists often use isotopes to distinguish the history, origin, and sometimes the age of various rocks. The amounts of some isotopes, such as carbon-14 and beryllium-10, are changed by the amount of cosmic radiation entering our atmosphere, which varies depending on the intensity of Earth's magnetic field. (Cosmic radiation is further discussed in chapter 6.) Collaborations between scientists from different scientific disciplines are needed if these interface problems are to be solved.

*   *   *

Following auroras, lightning is probably nature's second most spectacular light display. Many of you are aware of Benjamin Franklin's experiment with a kite in 1752 in which sparks jumped from a key attached to a kite string to his hand, showing that thunderstorms are electrified. But did you know that a couple of years earlier Franklin (1706–1790) had proposed that a man hold a 20- to 30-foot-long metal rod vertically upward from an elevated and insulated platform to test for electric effects? When a thundercloud moved overhead, the man was to move the rod toward a wire connected to the ground. If sparks jumped between the rod and wire, it would demonstrate that the cloud was electrified. Thomas-François d'Alibard (1703–1779) successfully carried out this experiment in France shortly before Franklin carried out his famous kite experiment. The soldier used by d'Alibard in this experiment was lucky that lightning had not occurred,[18] since the insulated platform probably would not have offered sufficient protection.

When I was a boy, my father explained how I could estimate the distance to a lightning strike by counting 1,001, 1,002, and so on after I saw a lightning flash. Each number takes about a second to say. Because light travels much faster than sound, I could estimate the distance in feet to the lightning by multiplying the number of seconds between the lightning flash and the thunder by 1,000. Five seconds meant the strike occurred approximately a mile away. However, at that time it was not recognized how complex an apparent "single" strike of lightning was.

Typically several lightning strikes are involved in what is visualized as a "single" strike. The first strike is usually from the ground to the cloud, contrary to popular conception. However, lightning is quite varied and includes strikes between clouds and even within the same cloud. Lightning is a bit like a huge capacitor. One plate is in the cloud and the other is on the ground. If the electric charge buildup is large enough, a discharge occurs in the form of a lightning bolt. It is also sometimes compared to the shock that occurs after you shuffle across a rug and touch someone—only on a much larger scale.

My father's method for estimating distance works because lightning and thunder occur essentially at the same time. The heating of the air associated with a lightning strike is often between 20,000°C and 30,000°C. This causes the air to expand suddenly and then collapse, which quickly leads to a shock wave we identify as thunder. Thunder usually is heard as a sharp boom, followed by a rumble. The rumbling occurs because the sound waves from the boom reflect off various objects such as topography, buildings, and trees. The closer you are to the strike, the sharper the boom is.

There are both positive and negative charges in a thundercloud (cumulonimbus). The mechanism by which electric charges build up in clouds involves the exchange of electric charges between ice crystals when they collide. This exchange results in smaller ice crystals having a positive charge relative to large ice crystals.[19] Updrafts within the cloud lift the smaller positively charged ice crystals relative to the larger negatively charged ones. On average, the top of thunderclouds are positively charged and the bottoms are negatively charged.[20] You should not expect to observe lightning from warm clouds lacking ice crystals.

When a significant negative electric charge builds up at the bottom of a cloud, it starts to jump toward the ground in a series of steps referred to as stepped leaders. Each step is about 50 meters long or about half the length of a football field. The leader may branch into many paths as it descends, and it provides a conducting channel along which electricity can readily flow. As the leader advances, opposite charges are induced on the ground. These charges rise to the top of any protruding objects that conduct electricity. I experienced this once while climbing along a ridge. I felt static charge pulling my hair up. I was one of the protruding objects. The leader descending from the cloud was probably no more than a few hundred meters above me. It was fortunate that I was not hit by lightning, as this was a dangerous situation. (Although escaping death, it took some time to recover my metal ice ax, which I had thrown off the ridge when I experienced the static charge buildup in my hair.)

After the leader gets within 10 to 100 meters of the ground, a positive charge moves up from the ground to close the gap between the ground and the leader descending from the cloud. This process establishes an electrically conducting channel, along which a so-called breakdown wave travels upward—the first electric strike. This strike moves upward from the ground to the air in about 100 microseconds (a hundred-millionth of a second).[21] This is immediately followed by a return strike carrying electrons to the ground. This return strike may, or may not, be the most luminous part of lightning. It is only an illusion that the first luminous part of the strike initiates at the cloud and travels downward to the ground. However, conditions under which lightning occurs are variable. For example, although the majority of the strikes in Washington state act this way, about 10 percent do not. Most exceptions carry an initial positive charge to the ground rather than a negative one. Some of the most deadly lightning strikes are between the ground and the anvil-shaped part of the cloud that extends outward from the top of the thunderhead. The variability of lightning reflects many factors, including the complex flow of air within thunderclouds, which can lead to an unusual charge distribution.

Lightning might be responsible for the gap in the Van Allen radiation belts. This gap, often referred to as the safe zone because of the low level of radiation present, lies between the altitudes of 10,000 and 13,000 km. It is known that lightning emits signals in the radio frequency range. You experience these as static on the radio following a lightning bolt. The radio frequency electromagnetic waves emitted by lightning are hypothesized to reduce the energy (which is proportional to the square of the particle's velocity) of particles between 10,000 and 13,000 km altitude. The particles then fall out of the safe zone into the lower Van Allen belt. If confirmed, this would solve one of the many mysteries in space science.

Lightning can affect space weather in yet another way. In 1975 a NASA test pilot flying a surveillance aircraft at a height around 20 km saw lightning moving upward to altitudes above him. This sighting of lightning remained questionable until Minnesota scientists photographed an upward-directed discharge in 1989. Before this, most scientists thought lightning was confined to the bottom of our atmosphere. During the past two decades, numerous lightning-like flashes have been observed to travel up into the ionosphere. Some of these have been seen even from space and are now under intensive investigation by scientists.[22]

Magnetic Orientation and Navigation by Animals

I remember as a boy in Michigan seeing large flocks of birds flying south for the winter. My parents explained they would come back in the spring. Some animals, such as birds, migrate for climatic reasons, while others migrate to reproduce. As a boy I didn't waste much time thinking about this puzzle. It seemed reasonable that animals would want to leave the cold winters of Michigan to fly to places like Florida. I don't recall wondering how animals knew which direction south was. Later I learned of the long journeys of salmon that ended with their death while spawning in the rivers where they were born. While curious about how this worked, I never imagined early in my career that I would become involved in research on animals that sense magnetic fields, the subject of this chapter, but I did.

Nearly three decades ago I reluctantly agreed to serve on the PhD committee of Tom Quinn, a fisheries graduate student at the University of Washington. I was initially wary of Quinn's proposed research, even though Ernie Brannon, a well-respected salmon expert, was to chair his PhD committee. Quinn proposed to determine whether sockeye salmon (*Oncorhynchus nerka*), one of five species of salmon in western North America, could sense Earth's magnetic field and use it for orientation. I advised him this was too risky: he might spend years carrying out research that would prove to be unpublishable. Then he would not receive a PhD degree. However, Quinn drew me into his program by telling me about some curious behavior of salmon.

Sockeye salmon fry (newly born fish) spend their first year in a lake before they go to sea to carry out a one- to four-year migration in the Pacific Ocean.[1] The fry instinctively know whether the lake is upstream or downstream from their hatching location. Apparently, this knowledge is genetically imprinted, as Ernie Brannon had demonstrated in his PhD research. Quinn also explained that sockeye salmon primarily identify the stream in which they are born by using an olfactory sense: they smell the impurities in the water.

Quinn wanted me on his committee to teach him about Earth's magnetic field and to suggest how best to control the local magnetic field in his experiments. As it turned out, I eventually learned much more from Quinn than he did from me. His proposed research turned out to be a success story. Quinn, now a full professor at the University of Washington, is still involved in salmon research, including how salmon impact the coastal ecology of Alaska.

Quinn used a simple but clever device to determine the direction salmon fry swim. He thought the majority of fry would swim in the migration direction even in a tank of water. Quinn's experiment involved releasing fry into a tank, after which they would swim into one of four arms containing traps making it impossible for them to reverse course. Electric coils surrounded the tank, which allowed Quinn to control the direction of the magnetic field (fig. 5.1). He constructed many traps and tanks and measured thousands of salmon fry. As he predicted, the newly hatched fry in the tanks swam in the same direction that they had been traveling in the river, and the direction that they would continue to travel in the lake. They did this even though their olfactory sense was useless in the tank environment because there was no flowing water. Quinn then changed the direction of the magnetic field by 90° to see if newly hatched fry would respond to the change in the magnetic field direction. They didn't. The experiments appeared to confirm my initial fears. But Quinn was nowhere close to being finished.

Quinn had carried out the experiments during the daytime. Light is transmitted as an electromagnetic wave that is polarized (chap. 1). You can determine the direction the polarized sunlight has by rotating polarized sunglasses by 90°. Perhaps salmon are capable of detecting this polarization. Or maybe salmon are using some other visual clue. To test the visual clue possibility, Quinn repeated his experiments at night. He also conducted experiments during the day using translucent covers to obscure the position of the Sun and distort polarized light. He also kept track of

FIGURE 5.1 One of Quinn's four-armed salmon tanks lays within a coil system. In this case the coil system—consisting of wires (shown) across the top, bottom, and on two sides—was used to control the direction of the magnetic field. The central bucket in the trap was used to lower salmon fry into the system. Photo from Tom Quinn.

which trials were conducted under sunny and overcast conditions. Those salmon not seeing polarized sunlight shifted their swimming direction by 90° whenever Quinn changed the magnetic field by 90°. Salmon could sense the magnetic field!

Quinn also showed that salmon did not use the dip of the magnetic field (magnetic inclination), but only the declination of the magnetic field—the same thing humans do when using a magnetic compass. Biologists often refer to this as sensing the "polarity" of the field. In contrast, physical scientists refer to polarity in terms of a complete flip of a magnetic vector—not one of its components, such as the horizontal component. For example, a polarity reversal of Earth's magnetic field involves a change in both the declination and inclination (chap. 2). Therefore I will refer to animals as "sensing the declination" when only the horizontal component of Earth's magnetic field is involved. As we shall see later, not all animals use declination, the deviation of magnetic north from true north, to sense Earth's magnetic field.

Evidently salmon, like humans, sometimes use more than one sense. When we cross a road, we not only look, but we listen, to see if traffic is approaching. Salmon use a hierarchy of senses in a river. The most important is smell, followed by sunlight, and then the magnetic field. Because the first two senses did not provide useful information when the tank was covered, the fry used the magnetic field to determine which direction to swim. They evidently do this at night in the natural environment.

Two questions occurred to Quinn. What mechanism do salmon use to sense a magnetic field, and do adults utilize Earth's magnetic field in the open-ocean migration phase? After completing his PhD degree, Quinn was able to find a partial answer to the first question. But he was not able to answer the second question, which still remains a puzzle.

However, before we consider these questions, it is useful to discuss what biologists mean when they refer to animal navigation and to introduce the so-called map problem. Navigation refers to the use of some mechanism by an animal to recognize its position with respect to a goal (some location it wants to reach) that cannot be directly detected. This is distinguished from orientation, which is the ability to maintain a constant course (e.g., with a compass). Because salmon can "directly" detect their goal by smelling the odors of plants, rocks, soil, and other chemical features in a stream, the use of the olfactory sense in a stream would not be considered a true navigation mechanism.

It is not clear how, or if, sockeye salmon use their olfactory sense for navigation in the open ocean, since they often use different outward migration routes than return ones. It has been suggested that salmon and other animals might sense the linear magnetic anomalies that record the reversals of Earth's magnetic field (chap. 2). However, the direction of these stripes is considerably different off the west coast of Washington and southern British Columbia from the Gulf of Alaska. How do salmon know that the anomalies change direction along their migration route? Other speculative methods of navigation run into similar problems and indicate that salmon have some "map sense." If you were placed in some unknown location, you could not use a magnetic compass to travel in the direction of the equator unless you were told that you were in the Southern or Northern Hemisphere. It would be even more difficult to go straight west on a cloudy day, even if you had a compass. You need some sort of map telling you how the declination varies. Figure 5.2 shows the variation in magnetic declination across the globe for a 2000 IGRF (chap. 1). Similarly, animals that migrate long distances also require more information than can be obtained from a compass. For example, if salmon use Earth's magnetic field in navigation, they need to know the correct declination at a given locality.

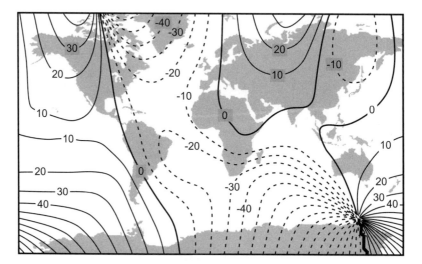

FIGURE 5.2 Declination in 10° contour intervals for the 2000 IGRF. Figure redrawn by Beth Tully from NASA images.

We will find that the map problem remains unsolved for animals that sense Earth's magnetic field.

Quinn was aware of evidence (to be discussed) showing that some animals sense the magnetic field by utilizing small grains of magnetite contained within their bodies. At the time Quinn was carrying out his PhD research, I was director of a large magnetic laboratory, which contained a superconducting magnetometer. This magnetometer used liquid helium to cool quantum mechanical sensing devices to 4° above absolute zero (−273°C). The magnetometer was so sensitive, it was difficult to obtain sample holders that would not produce a signal. Quinn used this magnetometer to conclude that salmon fry produced no magnetic signal. During one of our discussions, I learned that Quinn measured salmon fry that had been preserved with formalin, a solution of formaldehyde. I wondered what formalin did to magnetite. To find out, Quinn placed magnetite crystals in formaldehyde: they rapidly dissolved. He would have to repeat the experiments using fresh fry; ones caught the same day as they were measured. By this time some of my colleagues were becoming discontent. They were certain we "would lose one of those #@*&$ fish in the magnetometer and the magnetometer would smell like @#&*# fish into the foreseeable future." However, I assured them I would be present when Quinn made the measurements to make sure nothing went wrong.

I also said we would make the measurements after midnight when no one else wanted to use the magnetometer. I am not certain this pacified any of my colleagues, but they had little choice in the matter since I was then the director of the laboratory. This time around we detected a magnetic signal, but it appeared to come from the digestive tracts of the fresh fish. We concluded that the magnetometer was so sensitive it was picking up a magnetic signal from material the fry had consumed. This magnetic material should be viewed as a contaminant rather than some sensory device, a magnetoreceptor.

* * *

After Quinn completed his PhD degree, I decided I should be able to figure out the mechanism problem. After all, I was in the process of transitioning from an experimentalist to a theorist. Surely a theorist in physical science could solve a simple problem in biology. After thinking about this for a while, I came up with the following three classes of mechanisms:

1. Induced magnetization (see chap. 1).
2. Permanent magnetization; for example, permanently magnetized magnetite (Fe_3O_4) particles.
3. Electric currents induced by an animal moving in a magnetic field.

Armed with this information, I approached a biochemist. After patiently listening to me while I explained why the amount of electrical current traveling down a nerve might be altered by a magnetic field that changed relative to an animal, she gave me an elementary tutorial on biology. It turns out I was not as clever as I thought I was. Biology is a difficult subject, at least for this geophysicist. It was a lesson I should have learned at Berkeley during my graduate student years. There, a famous solid-state physicist, Charles Kittel, told me he was taking a sabbatical year to Europe to learn biophysics. He returned after three months to declare that biology was too difficult and that he was going to concentrate on physics. The only thing I can say on my behalf is it took me far less time to learn this lesson than it did Kittel.

What we call a nerve or neuron is a complex structure that includes the cell body, the axon, and the branching dendrites at the neuron's ends. The electric signal transmitted along the nerve is also not simple. Electrons do not transmit it, because the longest part of the nerve, the axon, is a poor conductor of electricity. Instead, the signal travels along the axon as an action potential—a voltage spike produced by ions diffusing across the cell membrane of the axon. Both potassium and sodium ions have a single positive charge. They move through so-called ion channels. When an axon is activated, potassium and sodium ions flow across the axon wall—perpendicular to the direction of the signal along a neuron. For every three sodium ions transmitted outward through an ion channel, two potassium ions are transferred inward through a neighboring ion channel. The outer part of the axon becomes positively charged relative to the inner part. There is a voltage difference resembling that between the poles of a chemical battery. Almost immediately after this happens, gates for the sodium ion channels close and ones for the potassium ions open, allowing potassium ions to neutralize the charge difference between the inside and the outside of the axon. The net effect of the ion flow is the production of an electric spike that lasts a few milliseconds at any one location. This spike is the nerve signal, called the action potential, that travels along a neuron until it reaches a synapse, a gap between neurons. There, a chemical is released, a neurotransmitter, that carries the signal across a gap of only 20 to 40 nanometers (20 or 30 billionths of a meter). There are many tens of known neurotransmitters, chemicals that relay and alter signals between neurons. Different neurons emit different neurotransmitters, which can be received by neurons in the vicinity of the releasing cell.

The magnetic field associated with the electric spike along a neuron is also complex. A dipole magnetic field is produced by a current flowing in a circular wire. However, a quadrupole field (chap. 1) is produced in a circular wire by an electric spike, modeled as a positive charge closely followed by a negative charge. The biochemist I consulted asked me if I had thought about mechanisms by which ion channels might have their properties altered by Earth's magnetic field? I had not. After my brief discussion with the biochemist, I decided to add a fourth class of mechanisms to my list:

4. A chemical reaction affected by a magnetic field.

Although I never published this list, the four classes of mechanisms given above are the ones actively being investigated by scientists. I will discuss these mechanisms further as we proceed to learn how animals, ranging from bacteria to mammals, sense Earth's magnetic field.

Nearly three decades ago I stopped actively carrying out research in the field after Quinn completed his PhD degree. I concluded I would have to learn too much biology to make a contribution to this subject. Quinn later worked briefly in Joe Kirschvink's laboratory at Caltech to confirm that adult salmon contain small magnetite crystals in their heads (mechanism 2 above).[2] This has also been found in several other species of fish, including tuna and trout. Quinn and Kirschvink concluded that salmon probably used these crystals to detect Earth's magnetic field, but this was not conclusively demonstrated. They were aware of Richard Blakemore's pioneering study in 1975 showing certain bacteria used chains of magnetite crystals to sense Earth's magnetic field, but the mechanism linking magnetite to the sensory physiology of higher organisms was not obvious.

*   *   *

In the late 1970s, Joe Kirschvink was in my office peering through an optical microscope at bacteria he had collected in the mudflats of Lake Washington bordering the University of Washington campus. He was attending a scientific meeting on campus, and he wanted to demonstrate that magnetotactic bacteria could sense the magnetic field. While I looked through the microscope, Kirschvink reversed a strong hand magnet next to the microscope. The bacteria immediately reversed course and began to swim in the opposite direction. These bacteria were sensing the magnetic field!

Richard Blakemore first reported this behavior in the journal *Science* in 1975. Blakemore and his colleagues, particularly Richard Frankel, even showed that magnetotactic bacteria swam, using a flagella (a tail-like appendage), in the opposite direction in the Southern Hemisphere (New Zealand) than in the Northern Hemisphere. (Earth's magnetic field has an upward component in the Southern Hemisphere and a downward component in the Northern Hemisphere; see fig. 1.1.)

In the muddy environments in which they live, the bacteria cannot use light. Nor can they use gravity because they are neutrally buoyant. These magnetotactic bacteria need to swim to the boundary between oxygen-rich and oxygen-starved water where they thrive.[3] The bacteria swim along the direction of the magnetic field lines to locate this boundary; if the oxygen level is too high, the bacteria die. When oxygen levels become too great, the bacteria reverse course and swim in the opposite direction. In the case of the simple experiment shown to me by Kirschvink, the chemistry of the bacteria's environment was invariant. Under those conditions, when the magnetic field was reversed, the bacteria reversed direction. This reversal was so clear that no statistics were needed to convince me that magnetotactic bacteria use the magnetic field to provide a direction in which to swim. But how do they do this?

Scientists, including Frankel and Kirschvink, have shown a magneto-tactic bacterium contains a magnetosome, a long chain of several magnetite (Fe_3O_4) crystals separated from each other by a thin layer of organic material (fig. 5.3). The submicron magnetite crystals are mostly uniformly magnetized ones, called single-domain grains (appendix), for which the magnetization is most stable. Smaller or larger crystals have less ideal magnetic properties. For an equal-dimensional single-domain crystal of magnetite, a change in only one-tenth of a millionth of a meter in the dimension of their sides makes them dramatically less stable to externally applied fields (appendix). Some scientists, such as Bruce Moskowitz at the University of Minnesota, have even used magnetosome crystals to study fundamental properties of magnetism because of the crystal's ideal magnetic properties. In 1990 Stephen Mann and associates (including Frankel) showed that some types of magnetotactic bacteria have magnetosomes consisting of greigite (Fe_3S_4), which also have ideal magnetic properties. Magnetosomes predominantly, but not exclusively, consist of single-domain magnetite crystals.[4]

Kirschvink emphasized in his discussions with me that the presence of single-domain magnetite in magnetosomes shows the bacteria are using them to sense the magnetic field: "Why else would they make magnetite crystals in such a narrow size range?"—a good question often posed by Kirschvink. His intuition has been strongly supported by experiments. In one experiment, the magnetization direction in the crystals was reversed by applying a strong magnetic field for a short time parallel to the long axis of the magnetosome. This is sometimes referred to as the pulsed magnetic field treatment. After this was done, the bacteria swam in the opposite direction. In another experiment, genetically modified bacteria lacking magnetosomes were produced. Because these bacteria could no longer sense Earth's magnetic field, they fared poorer than did bacteria with magnetosomes. Even though it takes energy to produce magneto-somes, magnetotactic bacteria gain an evolutionary advantage by doing so: magnetosomes can be used to find the optimum position in the mud for the bacteria to prosper.

Of all the animals that sense Earth's magnetic field, magnetotactic bacteria are the ones in which the mechanism is best understood. It is well established that they use the magnetite crystals in magnetosomes as magnetoreceptors, an example of mechanism 2 given earlier. A magnetic field will provide a torque on the magnetite. In response, the crystal will

FIGURE 5.3 A drawing of a transmission electron microscope photo of a magnetotactic bacterium. The black dots inside the bacterium are the magnetite crystals that constitute the magnetosome. Original photo provided by Richard Frankel, Department of Physics, Cal Poly State University, San Luis Obispo, California. Drawing of photo by Beth Tully.

attempt to rotate, which presumably is sensed by a bacterium by some mechanism that has yet to be clarified.

*   *   *

Animals of all sizes seem to sense Earth's magnetic field. Not only do single-cell animals such as bacteria sense Earth's magnetic fields; so do one of our planet's largest predators, the great white sharks.

The 5-meter-long great white shark rose up under the side of our boat and displaced it sideways. No, this was not another film based on Peter Benchley's book *Jaws*. It was an experience our family had in 2005 while cage diving with great white sharks off Seal Island in False Bay of South Africa. We were one of three boats that day witnessing the complete breaching of great white sharks as they shot up from a depth of 20 meters to attack seals swimming near South Africa's Cape Peninsula in one of the world's most dangerous, shark-infested waters. A second boat was carrying out research, while the final boat was filming for a nature show.

During our trip I learned about the extraordinary journey of one young female great white shark, informally called Nicole (after the Australian actress Nicole Kidman because of her apparent love of sharks) by the researchers. I followed up on this research after I returned to the University of Washington. A pop-up tag was attached to Nicole on November 7, 2003. This tag recorded temperature and pressure (which gives depth) approximately every minute and provided information on sunrise and sunset times. As programmed, the pop-up tag released itself from Nicole on February 28, 2004, rising to the surface, where it relayed information to a satellite. Researchers were excited to find the pop-up site was 37 km (23 miles) south of Exmouth Gulf in Western Australia. This shark had traveled at an average speed of 4.7 km per hour (nearly 3 miles per hour) over a distance of 11,000 km in about three and a half months. Two-thirds of the time Nicole swam within 5 meters of the surface, leading to the speculation that Nicole was using some visual clue for navigation, as is sometimes done by salmon. For reasons not understood, Nicole often took deep dives to depths sometimes exceeding 500 meters. The course Nicole took to Australia was crudely estimated based on latitude estimates obtained from the length-of-day information. It was a remarkably straight course, consistent with the high speeds estimated for the transit; any longer path would imply that Nicole swam even faster than estimated.

Back off the Cape of South Africa, researchers were continuing their studies of great white sharks. One component of these studies involves photographing the sharks' dorsal fins, which have distinctive marks useful for identification. On August 20, 2004, Nicole was photographed in False Bay. She had returned to her tagging site! Because there was no tagging information after February, no one knows how long Nicole spent off the coast of Australia or why she even went there. It is also not known how long Nicole had been back off the Cape of South Africa before she was photographed. Regardless, in nine and a half months or less, Nicole had made a round-trip to the west coast of Australia and back. Clearly, Nicole employed a precise way of navigating to have made this remarkable journey.[5]

A paper published in 2004 in *Nature* by Carl Mayer and his colleagues in Hawaii demonstrated that sharks could be conditioned to seek food by swimming toward a magnetic field produced by a coil system at the side of a 7-meter-diameter tank. These scientists hypothesized that the sharks sense magnetic fields by using specialized electrically conducting channels in their heads.

Elasmobranches are a subclass of fish—primarily consisting of sharks, skates, and rays—with stiff cartilage rather than bone for structural support. These fish have hundreds to thousands of electrically conducting canals in their heads (depending on the species and individual) referred to as ampullae of Lorenzini (fig. 5.4). The ampullae are jelly-filled canals that are remarkably sensitive to electric fields. Indeed, sharks are often described as the animal most sensitive to electric fields: they can sense voltage differences as low as five one-billionth of a volt per centimeter. This is equivalent to detecting the voltage of a three-volt flashlight battery with poles separated by 6,000 km (3,720 miles). Naturally, they don't sense electromagnetic signals from great distances because there are many close-by sources, including those from other animals and from induced electromagnetic signals from changes in ionospheric magnetic fields (chap. 4).

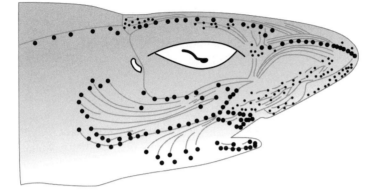

FIGURE 5.4 There are many electroreceptors, called the ampullae of Lorenzini, in the heads of sharks and rays. These are shown above as black dots. Figure redrawn by Beth Tully from *Wikimedia Commons*.

Sharks use their ampullae to sense electromagnetic signals emitted from prey. For example, small sharks often prey on flounders, flat fish with two eyes on one side of their heads that often bury themselves in a thin layer of sand on the ocean bottom. In 1971 Adrianus J. Kalmijn, a student at the University of Utrecht in the Netherlands, published his PhD research showing that sharks (*Scyliorhinus canicula*) attacked electrodes buried in the sand that mimicked the muscle contractions of flounders. Further experiments by Kalmijn and others have confirmed that the prime purpose of the ampullae is to locate prey.

The possible use of the ampullae of Lorenzini to sense Earth's magnetic field is an example of mechanism 3 given earlier.[6] The ampullae are tubes that end as pores on the skin's surface; they are directly connected to the electrically conducting seawater. As discussed in chapter 3, an electromagnetic generator works because an electric voltage is produced when a wire is moved at right angles through a magnetic field. A similar mechanism to generate electricity may be employed by sharks. Although somewhat oversimplified, an electrically conducting loop (analogous to a wire) can be envisaged as attached to a shark's head. It consists of the ampullae inside and the seawater outside the shark's head. When a shark, or any other elasmobranch, moves its head—say, from side to side as sharks are often observed to do—an electric current is induced in this moving loop by the magnetic field.

In theory, these induced currents in the ampullae could be used for navigation. However, in practice, I am not aware of any convincing evidence that sharks actually do this. Indeed, Kirschvink and his colleagues, especially Michael Walker, argue that sharks sense the magnetic field the same way salmon do—by using magnetite magnetoreceptors in their heads. They argue that this is more effective than using the ampullae of Lorenzini to do both, which they say could make catching prey less efficient. In 2003 Walker, Carol Diebel, and Kirschvink trained stingrays (elasmobranches) to sense a local magnetic field. They then inserted magnets up some stingrays' nostrils and brass weights up the nostrils of others. The stingrays were more affected by the magnets than the brass. Walker and his colleagues argue that this shows that the rays use magnetic particles in their heads, which would be affected by the magnets. In contrast, they argue, the ampullae would not be affected because their position relative to the magnets remains fixed and hence should not induce any potential differences. I am not completely convinced by this because the movement of stingrays containing magnets could induce electric currents in the seawater near the stingrays, which then could be sensed by the ampullae.

I conclude that the magnetoreceptor problem in elasmobranches has not been solved. Elasmobranches can sense magnetic fields. They can even be conditioned to magnetic fields. However, no definitive evidence indicates that sharks use Earth's magnetic field for navigation, even though the long migrations of some sharks, such as Nicole, make this an attractive hypothesis.

*   *   *

There are several other ocean travelers besides salmon and sharks that seem to employ sophisticated navigation schemes, as suggested over a half a century ago by Archie Carr (1909–1987). Carr, a University of Florida professor, was a naturalist who became one of the world's foremost authorities on sea turtles in his time. An anecdote included in his 1956 book, *The Windward Roar: Adventures of a Naturalist on Remote Caribbean Shores*, suggests that green (sea) turtles (*Chelonia mydas*) have a remarkable navigation system. A captain on a fishing boat related the capture of a green turtle off Nicaragua. The fisherman who caught it inscribed his initials on its shell, a common practice at the time. The turtle escaped death when the boat capsized off the Florida Keys in a violent storm. Months later the turtle was recaptured in the same area off Nicaragua where it had been originally caught, about 1,000 km from the Florida Keys.

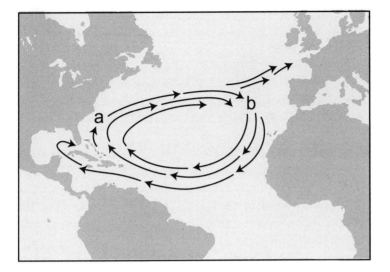

FIGURE 5.5 Loggerhead turtle migration begins in the southeastern United States, travels in a northerly route across the Atlantic from points *a* to *b*, and then back on the southerly route from *b* to *a*. Figure redrawn by Beth Tully from original provided by Ken Lohmann.

Fortunately, we do not have to depend on anecdotes, which can be noto-riously unreliable, thanks to the efforts of scientists such as Ken Lohmann and his colleagues (often including his wife, Catherine) at the University of North Carolina, who have shown many seagoing animals—including turtles, lobsters, and nudibranchs (sea slugs)—sense Earth's magnetic field. Although accounts of their research are of wide interest, including their work on green turtles, I will only briefly summarize some of their findings to illustrate that different species of animals sense different ele-ments of Earth's magnetic field.

Loggerhead turtles (*Caretta caretta*) are large turtles that can grow to more than a meter in length, weigh up to 360 kg (792 pounds), and can live more than 190 years. As do all seagoing turtles, loggerhead females return to the same beach to lay their eggs, having been impregnated some-where along a long migration route. Hatchlings born on Florida's east coast beaches instinctively swim out across the waves to make one of the longest and most spectacular marine navigations. They swim out along the Gulf Stream to the North Atlantic Gyre, which encircles the Sargasso Sea (fig. 5.5). Although this gyre provides young turtles with abundant food, they must not stray too far northward without significantly risking death. Somehow these turtles appear to know when they are too far north, as most swim southward when this happens. Young turtles often spend several years near the Sargasso Sea before continuing on with their migration. As shown in figure 5.5, the outgoing migration route differs from the return route. This is also the case for many species of salmon. How do the turtles know which way to swim? Research indicates that they have a directional compass programmed into their DNA telling them which way to go.

Ken Lohmann and his colleagues captured newly hatched turtles and used coil systems similar to, but larger than, those Quinn used to control the magnetic field (see fig. 5.1). They used three different magnetic fields: each represented a different location along the migration route. Hatchlings exposed to a magnetic field similar to the one at point a on figure 5.5 preferentially swam in a north–northeast direction. Those exposed to the magnetic field they would experience at point b on figure 5.5 swam southward. Finally, those exposed to a magnetic field characteristic of the southernmost part of the turtle's expected migration path swam eastward. The evidence indicates that turtles instinctively use Earth's magnetic field to swim in the appropriate migration direction.[7]

A fundamental difference from salmon emerged during these experiments. Lohmann and his colleagues showed that loggerheads sense magnetic inclination and magnetic intensity, but not magnetic declination. Salmon sense declination and intensity, but not inclination. Three magnetic elements—inclination, declination, and intensity—are used to characterize Earth's magnetic field. They vary from location to location because the dipole magnetic field is tilted with respect to the rotation axis and because a significant nondipole field exists (chap. 1). Why sockeye salmon have evolved to sense declination, while loggerhead turtles have evolved to sense inclination, is a mystery. Turtles also need a map sense to navigate, but how this is done is not fully understood.

Although other marine animals make shorter journeys than turtles, many of them, including lobsters, also appear able to sense Earth's magnetic field. For example, lobsters, which are invertebrates (that is, they lack backbones), seem to exhibit "true navigation" ability in that they determine their position without relying on information about the surroundings acquired during their travels. Spiny lobsters (*Panulirus argus*) often make long nocturnal foraging trips. They also make even longer annual migration trips. Near Bermuda, these spiny lobsters migrate long distances to form aggregations off the Bermuda shelf.

It is not clear why they do this. An exodus from areas around Bermuda occurs after the first autumn storm, suggesting to some scientists that the lobsters are migrating to deeper waters to move away from colder shallow water conditions that occur near Bermuda in winter. Other hypotheses include that they migrate for molting and mating purposes. Lobsters form long lines during these migrations. Sometimes tens of individuals are in tactile contact through their antennae when they make this journey. It has been speculated they do this to minimize the energy expended during migration by reducing drag produced by ocean currents, much like some bicyclists do in a race. In any case, it has been known since the middle of the previous century that lobsters (tagged for identification purposes) can return "home" after being displaced by more than 20 km (12 miles).

During this century, Lohmann and his colleagues have shown that the spiny lobsters appear to sense Earth's magnetic field, based on experiments using electric coils to control the magnetic field and observing the directions lobsters prefer to move. Lobsters, like many other animals that migrate, including salmon and turtles, appear to sense Earth's magnetic field. They have also solved the map problem, but scientists remain in the dark as to how they do this.

* * *

Biologists often turn to certain animal species when determining how some property, such as a magnetic sixth sense, is genetically recorded. One of those species is *Dropsophilia melanogaster*, which are disliked by almost everyone but biologists, who love them. *Dropsophiliae*, more commonly known as fruit flies in North America, are ubiquitous and annoying when fresh fruit is around. They are probably the animals most studied by biologists, especially geneticists: they are easy to grow and care for and can produce more than 800 eggs in their 30-day lifetimes at 28°C, the optimum temperature for survival. They also have only about 200,000 neurons in their heads and just four pairs of chromosomes, one of which is the sex chromosome.

A typical fruit fly is only about 3 mm long, has a yellow-brown color, and brilliant red eyes. Thomas Hunt Morgan (1866–1945) received a Nobel Prize in 1933 for his research on fruit flies—the first Nobel Prize awarded for genetic research. He showed that genes, the basis of heredity, were carried on chromosomes. The research had a serendipitous beginning. One day a white-eyed mutant male fruit fly appeared in the colony maintained in Morgan's "fly room" at Columbia University in New York. Morgan decided to mate this mutant with a red-eyed female. Only red-eyed offspring were produced, showing the mutation was recessive. However, when this generation subsequently mated among themselves, no white-eyed males appeared, much to Morgan's surprise. In contrast, the generation after this one exhibited a large number of white-eyed males (mixed in with red-eyed males, red-eyed females, and white-eyed females). Why did this occur? You may wish to pause here to work on this puzzle, before I provide you with Morgan's answer.

Morgan concluded that the gene for fruit fly eye color is carried only on the X chromosome, the first demonstration that properties other than sex are determined by the sex chromosome. Females have XX sex chromosomes while males have XY sex chromosomes. The mutant male could only have contributed a Y chromosome to males in the next generation. They all have red eyes, because none of them carried the mutant X gene (or, for the technically minded biologist, the allele), which could only have been passed on to females. However, those females in the following generation could pass on this mutant X gene (allele) to males.[8] As an aside, fruit fly research is presently at the center of a controversy concerning how important certain regulatory DNA elements are in evolution.[9]

Fruit flies like sugar. They like it so much they can be conditioned by a magnetic field to obtain it. This has been demonstrated by using a T-maze. The fruit flies are raised up in a vertical tube in an "elevator" to a level where a fork occurs. They must choose to move along a horizontal tube going left or going right. Coils are at the ends of both directions, but only one is turned on to generate a magnetic field. The fork containing the food is the one in which the coils are activated. Eventually the fruit flies become conditioned to associating a magnetic field with a food source: they would move in the direction of the magnetic field in later experiments, even when no food was present. But how do they sense the magnetic field?

Robert Gegear and Steven Reppert at the University of Massachusetts think they know the answer. In 2008 they reported in the journal *Nature* on experiments with fruit flies in a T-maze containing a light source. By using filters, they could control the wavelengths of the light used in the experiments. They found that the fruit flies could only sense the magnetic field when ultraviolet-blue light (wavelength between 300 and 420 nm) was present. They suggested that the magnetoreceptor involves cryptochrome (often referred to as CRY by geneticists), a protein that serves as a photoreceptor (senses light), present in the fly's eyes. A slightly different form of cryptochrome is used by a wide variety of plants and animals to control their circadian rhythm, the approximately 24-hour-long internal biological clock. This 24-hour biochemical cycle is used by plants to optimize growth and by corals to coordinate mass spawning. Cryptochrome is also known to affect the fly's circadian rhythm. Gegear and Reppert concluded that the cryptochrome serves dual purposes in fruit flies: one is for circadian rhythm, and the other is for magnetoreception.

A second paper, also published in *Nature* about the same time, supported this conclusion. The French scientist François Rouyer carried out similar experiments using mutant fruit flies that lack cryptochrome. These mutant flies could not be conditioned to the presence of a magnetic field, while those with cryptochrome could. The presence of cryptochrome appears necessary for fruit flies to sense a magnetic field.

But how does cryptochrome respond to a field? The answer is it forms a "radical pair," which can only be understood through a subfield of quantum chemistry called spin chemistry, a subject first developed during the 1970s. I will not try to explain this here, but I will give an oversimplified overview of the so-called radical-pair mechanism.

Many people consume vitamins C and E in a controversial attempt to reduce damage to their cells stemming from highly reactive "free radicals," electrically charged atomic groups. An example of a radical is the hydroxyl group (OH), which can be created from water (H_2O or H^+OH^-). This hydroxyl radical is highly reactive in many circumstances because it has a net minus electric charge. For example, the hydrogen ion, H^+, rapidly combines with the hydroxyl radical, OH^-, to form water.

In 2000 Thorsten Ritz and his colleagues at the University of Illinois suggested that incident light on the ends of some molecules could produce radicals. These radicals can have different magnetic strengths depending on how their electron spins combine.[10] (Electron spins can be thought of as producing tiny bar magnets, as we saw in chapter 1.) Providing the magnetic strengths of the radicals are different, they can affect the rate at which chemical reactions occur. Because two radicals are required, this is called the radical-pair mechanism.[11] Essentially it is a photochemical reaction (a reaction involving light) whose rate is affected by a magnetic field.

In 2008 two English chemists, Peter Hore and Christiane Timmel, carried out laboratory experiments on a complex molecule called CPF (for carotenoid, porphyrin, and fullerene), which is found in cryptochrome in the eyes of fruit flies (and in birds, as we shall see). Incident light produces radicals at the ends of CPF, which alters the way the molecule responds to a magnetic field. Hore and Timmel showed that a magnetic field ten times the strength of Earth's magnetic field affected the rate of certain chemical reactions involving CPF, consistent with the theory developed by Ritz and his colleagues. (A magnetic field larger than Earth's was used to make the effects easier to measure.)

At present scientists can only speculate on how animals sense the chemical reaction rate in cryptochrome. The long cryptochrome protein molecules, which occur in the eyes (retinas) of fruit flies, will generally be oriented in different directions with respect to the magnetic field. Some scientists speculate that the signal from the cryptochrome may "piggy-back" on the normal light signals sent from the retina to the brain. Depending on the orientation of the cryptochrome relative to the magnetic field, the light intensity could be increased or decreased. Because there are several cryptochrome molecules present with different orientations, this might, say, produce a pattern of dark and light spots. When a fruit fly changes its position with respect to the magnetic field, the pattern of spots moves across its field of vision. The animal uses the shift of the spots to sense the magnetic field. If this speculative model is verified, it essentially means some animals can "see" Earth's magnetic field.

The applicability of this mechanism (an example of mechanism 4 given earlier in this chapter) to animals is uncertain. Because of many unknowns, scientists have made several assumptions in the theory. There are also uncertainties regarding the applicability of experiments to real-life situations. Although the experiments carried out in a laboratory for the compound CPF show the mechanism is feasible, it has not been demonstrated that fruit flies actually use this mechanism. Nevertheless, the radical-pair mechanism is currently the best explanation for how cryptochrome might be used by fruit flies (and birds) to detect a magnetic field when ultraviolet-blue light is present.

Although other insects have been claimed to sense the magnetic field, sometimes these claims do not hold up. For example, studies of the monarch butterfly (*Danaus plexippus*) illustrate that some animals seem to make remarkable migrations without using a magnetic compass. Monarch butterflies east of the Rocky Mountains migrate in autumn to Mexico (Michoacán), while those west of the Rockies migrate to California (Pacific Grove and Santa Cruz). Because the duration of the migration far exceeds the lifetime of any one individual, the navigation direction appears to be programmed into their DNA. Usually after the egg, larvae (caterpillar), and pupal (chrysalis) stages, the monarch butterflies live for only 5 or 6 weeks. But after a few generations have come and gone, the autumn generation that initiates the migration has an expected lifetime of 6 to 8 months. If this variation in life expectancy applied to humans, some people would live for more than half of a millennium.

Because it was puzzling how some monarchs migrate hundreds to thousands of kilometers to some locality they have never seen, it was not surprising that a magnetic six-sense hypothesis arose. It is often difficult to eliminate false information once it is published in a respectable scientific journal. In 1999 a paper was published in the *Proceedings of the National Academy of Sciences* (*PNAS*) claiming that monarchs orient with respect to the magnetic field. Although this paper is still touted on the Internet, it is rarely mentioned that the authors published a retraction (in *PNAS*) several months later after they could not reproduce their earlier results. To my knowledge, there is no successful test showing that monarchs possess a magnetic compass, even though there is some evidence showing they contain magnetite in their thorax.

In 2002 two Canadians, Henrik Mouritsen and Barrie Frost, carried out some clever experiments with tethered monarchs under controlled laboratory conditions (also published in *PNAS*). Using a coil system to produce a magnetic field, they showed that monarchs did not change the migration direction when the field was rotated by 120°. Mouritsen and Frost then demonstrated that monarchs use a Sun compass by using the position of the Sun and their internal clocks to determine directions. (We use a Sun compass when we distinguish east from west at sunrise.) One group of wild-caught monarchs were placed in a controlled light/dark cycle for more than 5 days to reset their circadian rhythm clocks by 6 hours, while a control group did not have their 24-hour clocks reset. Flight directions were measured outdoors under sunny skies in a location where the monarchs could not view any geographical landmarks. For the latitude and dates of the experiments, the Sun's azimuth changed by slightly more than 90° over a 6-hour interval. The butterflies that had their clocks shifted by 6 hours changed their flight direction by about 90° relative to the control group. That is, monarchs appear to use a Sun compass, but not a magnetic one.

In 2008 neurobiologist Steven Reppert and his colleagues at the University of Massachusetts showed that monarchs use cryptochrome to do this. However, the cryptochrome used (called CRY2) differs from the one used by fruit flies (labeled CRY1). In summary, fruit flies use CRY1 for circadian rhythm and to sense Earth's magnetic field, while monarchs use

CRY2 to sense and link their daily clock to the Sun compass. Apparently at sometime in the distant past, fruit flies and monarchs evolved different types of cryptochrome to assist them in navigation.

Not all animals that sense a magnetic field do so to determine the proper direction for travel. While living in Australia, we found that termites have a profound effect on the landscape, particularly in northern Australia, where around two-thirds of Australia's estimated 263 termite species live. Unlike other colonizing insects—such as ants, bees, and wasps—termites do not undergo dramatic changes as they grow. Instead of changing from grub-like larvae to a pupal stage before maturing to adults, juveniles closely resemble small adults. Because of this, some scientists say termites are more closely related to cockroaches than to other colonizing insects. However, these "white ants," as our Australian friends call termites, also have many similarities to other colonizing insects. They have a worker caste making up the bulk of a colony and a soldier caste, whose individuals have large jaws for defending the colony. They also have a queen and king who found the colony after a winged mating flight.

Fortunately, I was able to travel to the Northern Territories of Australia, just south of Darwin, where I visited magnetic termite (*Amitermes laurensis*) mounds. Because during the summer monsoon season the region is often flooded, these termites build their homes aboveground. A mature mound is more than a meter in height (and sometimes nearly twice this height), as shown in figure 5.6. The tall thin mounds are aligned north–south, with their flat faces facing east and west. The king and queen produce eggs at the bottom of the mound, while most of the dead are buried near the top. These thin-skinned termites are sensitive to small changes in temperature. If the temperature deviates much from 30°C (86°F), they die.

But why are these mounds aligned north–south, unlike all termite mounds of other species? The answer has been known for sometime: they build their mounds to control temperature. If the mounds deviate substantially from a north–south alignment, the termites suffer. Heat from the Sun is absorbed by the east and west faces to keep the colony from becoming too cold. When the heat becomes too much on a scorching summer day, the termites move inward where it is cooler. Before a biologist at the University of Sydney, Peter Jacklyn, started to study termites, it was thought that these termites oriented their mounds by using the Sun. However, there is a serious problem with this hypothesis: the worker termites that build and repair the mounds are blind. Jacklyn carried out a simple

FIGURE 5.6 Magnetic termite mounds in the Northern Territory of Australia. The flattened sides of these mounds face east and west. They are about 1.5 meters tall, but only about 15 centimeters thick. Photograph by Ronald Merrill.

experiment. He broke down much of a termite mound and then inserted hand magnets into the remainder. Magnetic termite mounds consist of many small-elongated cells that are approximately aligned in a north–south direction. The damaged mound was repaired in a few months. However, the newly repaired cells near the magnets were no longer preferentially oriented in a north–south direction. Apparently they had been affected by the local magnetic field produced by the magnets.

If Peter Jacklyn's studies are confirmed by more research, it means "magnetic" termites build their mounds by using Earth's magnetic field. Moreover, since the workers are blind, they are not using cryptochrome to sense the magnetic field. They must employ some other mechanism. Perhaps it is the same mechanism that some bees use. James Gould, Kirschvink, and Ken Deffeyes conclude in a 1978 *Science* paper that bees sense Earth's magnetic field by using magnetite particles in their abdomens.

The Austrian scientist Karl von Frisch (1886–1982) received a Nobel Prize in 1973 for his work on honeybees (genus *Apis*). He showed that bees could not see red light (which we can see), but they could see ultraviolet light (which we cannot see). Most relevant to this discussion, he showed that bees transfer information to each other on the distance and direction of food sources by using a Sun compass, backed up by the ability to sense polarized light. This information is conveyed inside the hive by a waggle "dance." A bee moves along a line on the side of the hive while waggling her hind end. This line makes an angle with respect to the vertical that provides the direction other bees must fly with respect to the Sun to find a food source. At the end of the waggle part of the dance, the bee either turns left or right to circle back to the starting point of the dance. After repeating the waggle section again, she turns in the opposite direction from the one previously used to return to the starting point. This alternation between right and left turns produces a figure eight pattern. This pattern is sometimes repeated more than a hundred times. The duration of the dance is correlated with the distance to the food source. Bees that have spent some time in the hive even adjust the angle of the line along which they waggle to take into account the movement of the Sun, suggesting a biological clock is also involved.

Subsequently, it has been postulated that bees don't make the correct geometric conversion: errors of up to 20° often occur. From 1968 into the early 1970s, Martin Heisenberg and Martin Lindauer reported that these errors were systematic: the "errors" were conveying information. That is, the bees seem to agree on the misdirection information. For example, the majority of the bees carrying out the dance may agree that the angle is less than expected from only using a Sun compass. Heisenberg and Lindauer found the "error" vanished only when the line of the waggle portion of the dance was parallel to the magnetic field direction. The greater the line deviated from the local magnetic field direction, the larger the "error." They interpreted these results as showing that bees also use the magnetic field for navigation. They point out this would be useful on dark overcast days when the two primary mechanisms of navigation (the Sun compass and polarized light) were not available.

In papers published from 1978 to 1980, Gould and his colleagues produced data showing small superparamagnetic (appendix) magnetite crystals were made late in the pupal stage of development. They hypothesize that bees are using these particles, which are confined to a narrow size range, to sense Earth's magnetic field. Nevertheless, because the magnetic particles provide only indirect evidence and the claimed trends in the "errors" are small, I conclude that the evidence for bees sensing Earth's magnetic field is inconclusive and needs confirmation.

*   *   *

Many historians attribute the origin of the term "animal magnetism" to Franz Anton Mesmer (1734–1815), a Vienna physician who alleged that magnets could cure illness, only to find his ideas later debunked by a Paris Academy of Sciences committee, on which Benjamin Franklin served. Mesmer claimed magnetism traveled as a fluid that could remove obstacles in the body that produced illness. In spite of the claims of charlatans such as Mesmer, the first convincing scientific studies showing that magnetic fields could affect animals involved birds.

Bird migrations were recorded 3,000 years ago by natural philosophers including Homer and Aristotle. Although many birds are residents that don't migrate, many others do and some take incredibly long journeys. Bar-tailed godwits are believed to make the longest nonstop flight: in 2008 one was tracked to have flown 11,680 km (7,240 miles) nonstop from breeding grounds in Alaska to New Zealand. The godwit flew more than eight days without food, water, or rest. The arctic tern makes the longest-distance migration, including stops, of any animal: after breeding in the Arctic, it flies to its feeding locations in Antarctica. Even small birds such as warblers, hummingbirds, and flycatchers often migrate long distances. Usually they do this at night, apparently to reduce predation, or to use daylight hours to acquire fuel to fly long distances.

Scientists have long wondered how birds navigate such long distances, leading to speculations that they use Earth's magnetic field. As early as 1859, Alexander von Middendorf suggested that birds use the direction of Earth's magnetic field for navigation. Shortly thereafter, in 1862, C. Viguier suggested that birds use the intensity of Earth's magnetic field. (The intensity can be used to find the latitude, at least in theory, because it decreases by a factor around two between the poles and equator—as we saw in chapter 1.)

Birds use many environmental clues for navigation. Shortly after World War II, a German scientist, Gustav Kramer, pioneered modern navigational studies by showing that starlings tended to orient in their preferred migration direction in a circular cage, sometimes called a Kramer cage. He carried out the first experiments showing that some bird species use sunlight for orientation. He also was the first (in 1953) to suggest that birds need a map, as well as a magnetic compass, for navigation involving Earth's magnetic field. Later, a Cornell University scientist, Stephen

Emlen, showed that indigo buntings changed the direction they tried to fly in a Kramer cage when the star pattern was rotated in a planetarium. It is now recognized that birds use polarized sunlight, odors, local landmarks (such as rivers), stars, and, as we shall see, magnetic fields to navigate.

In 1958 the Dutch scientist A. C. Perdeck, was studying European starlings. These starlings migrated from Scandinavia and northwestern Russia to the southern British Isles. After catching and tagging the starlings in Holland, Perdeck transported them to Switzerland and released them. Experienced birds, ones who had made the migration before, flew at 90° to their usual course to arrive in the British Isles. In contrast, inexperienced birds, ones who had never made the journey before, flew in the direction that would have taken them to the British Isles had they not been displaced. They ended up in Spain and Portugal. The offspring of the birds that flew to Spain or Portugal continued to use the new migratory route. That is, they were spending their winters on the Iberian Peninsula rather than in the British Isles. The course these birds fly is coded in their DNA. This study convinced many ornithologists that some birds use a magnetic compass for orientation. But that is not the entire story. The experienced birds also use additional information they had learned from previous migrations.

A substantial number of studies during the twentieth century demonstrated that many species of birds use Earth's magnetic field for navigation. Homing pigeons proved to be particularly useful because they were selectively bred from rock pigeons (*Columba livia*) to return home after being displaced long distances. Scientists found that the direction traveled by homing pigeons was disrupted when magnets or small coils producing magnetic fields were mounted on their heads. In contrast, when dummy magnets or coils were used, the return route was not disrupted. Although substantial evidence now demonstrates that migratory birds use Earth's magnetic field for navigation, it is not clear how they do this.

The avian compass differs from the mechanical compass used by humans. Numerous scientists, particularly the American John Phillips and the German husband-and-wife team of Wolfgang and Roswitha Wiltschko (and their colleagues), have carried out experiments demonstrating that the avian compass depends on inclination and intensity (similar to turtles). The implications are many. Because a mechanical compass used by humans points northward, we have little difficulty in distinguishing between magnetic north and south when we are on the magnetic equator, where the inclination vanishes (chap. 1). However, birds apparently cannot do this with the avian compass because they do not use declination. It is not clear how birds migrating across the equator compensate for this. Perhaps they simply take into account various land features or the position of the Sun (or at night, the stars) when they stop to rest or feed. They might use these to continue on in the same general direction when they resume migration.

Debate on how birds sense Earth's magnetic field centers around the two mechanisms we have previously explored: the use of small magnetized grains in the head of birds and the use of cryptochrome as a photochemical magnetoreceptor. To understand the complexity of the debate, we first consider the evidence supporting the hypothesis that small magnetite particles serve as the magnetoreceptors. Subsequently, we examine the evidence supporting the cryptochrome hypothesis.

Based on her PhD research, New Zealand scientist Cordula Mora (and her colleagues) published a paper in 2004 convincing many scientists that the magnetoreceptor consists of small magnetite particles. She and her colleagues used food as a reward to condition homing pigeons to a magnetic field. The pigeons learned to walk to one end of a tunnel when a magnetic field was turned on and to the opposite end when it was off. In an attempt to distinguish between a photochemical magnetoreceptor and a magnetic particle receptor, they anesthetized the olfactory cavity, where small magnetic particles reside. The pigeons could no longer tell whether the magnetic field was on or off, even though their eyes were unaffected. Mora and her colleagues then carried out more detailed experiments by severing particular nerves in the olfactory cavity to show that the pigeons were not using smell in some way but were using a nerve argued to be activated by the magnetized particles.[12]

In 2002 Walker and Kirschvink and others proposed that the small magnetic particles in the beaks of birds were superparamagnetic ones (appendix), particles magnetized by an inducing magnetic field (chap. 1). As the direction of the inducing field changes, so does the direction of the magnetization in these particles. Kirschvink and his colleagues hypothesize that, in the presence of a magnetic field, these particles torque the nearby organic tissue to open some ion channels, while closing others. This elicits neural activity that might be sensed by the bird.

This mechanism is favored by another New Zealander, Todd Dennis and his colleagues, who reported in 2007 the results of releasing homing pigeons in the vicinity of a strong local magnetic field anomaly produced by magnetized crust (chap. 2). These pigeons were tracked using a small GPS system attached to their backs. Many birds had to fly to high altitudes, sometimes above 4 km, before they settled on the proper homing direction.

The scientists interpreted these results as showing that the pigeons flew to these altitudes to reduce the intensity of the local magnetic anomaly. (Local magnetic field anomalies are described by very high-degree spherical harmonics that decrease rapidly from the source; see chap. 1). Only at high altitudes could the pigeons discern Earth's main magnetic field clearly enough to return home. They argue this confirms that pigeons use both magnetic intensity and direction (inclination) information during navigation.

Advocates for a photochemical magnetoreceptor have argued that only circumstantial evidence supports the use of magnetic particles in beaks of birds. In contrast, or so they claim, the evidence showing that birds "see" the magnetic field is far more convincing. In humans, different photoreceptor cells are present in the eye in each of three different types of cone cells. Similar to the need for having three different primary colors to produce all the colors in a painting, the brain interprets what we "see" by integrating signals from cones, which allow for the perception of color, and the rods, which, although more sensitive to light, do not process color. Unlike humans, most primates can only sense two separate wavelengths: they do not see color as well as we do. In contrast, birds have four types of cones. They see more than we do. They see ultraviolet light, the light responsible for sunburn, in addition to the colors humans see.

Males in most bird species have colorful plumage used to attract mates and to defend their territories from rivals. Although the males of a few species appear drab to us, they are often brilliant when viewed under ultraviolet light. The bird's brains must integrate the separate wavelengths of light received by different types of cones, just as we do, to produce a complete picture. To my knowledge, no scientist has yet investigated the possibility that some animals, such as birds, use electromagnetic waves confined to the "non-visible" (to humans) range to sense Earth's magnetic field. (The fruit fly experiments included some visible light.)

Like fruit flies, birds have cryptochrome in their (right, but not left) eyes, which appears to be necessary for sensing the magnetic field.[13] However, different birds have slightly different forms of cryptochrome and appear to "see" the magnetic field in somewhat different ways. Experiments indicate that all songbirds so far tested (including European wrens, European robins, Australian silvereyes, and garden warblers) sense Earth's magnetic field when blue light (425–525 nm wavelengths) is present. When only longer wavelength light is present, these birds become disoriented, consistent with the radical-pair mechanism in cryptochrome, which predicts the sensitivity of cryptochrome to magnetic fields will disappear at long wavelengths. Wiltschko and his colleagues showed that although

European robins (*Erithacus rubecula*, which is a different species from the American robin), bobolinks (*Dolichonyx oryzivorus*), and homing pigeons could sense the magnetic field when blue light was present, they became disoriented under orange to red light (590–635 nm). Although this is consistent with the radical-pair mechanism, a problem has emerged. In 2004 Wiltschko and his colleagues showed that European robins, which were initially disoriented under red light, could be conditioned to sense the magnetic field when red light was present. Theoretical analysis suggests that the radical-pair hypothesis should not work at these longer wavelengths. Perhaps some other mechanism is operating to allow birds to sense Earth's magnetic field.

Interestingly, homing pigeons apparently require blue light to sense the magnetic field, yet they also cannot sense the magnetic field when the olfactory cavity is anesthetized. Assuming these results are reproducible (not yet tested), the brain appears to be integrating signals from the eye and the olfactory region to sense the magnetic field in some unknown way.

In a 2005 article in *Science*, William Cochran and his colleagues reported that Swainson's thrushes (*Catharus ustulatus*) use the sunset to calibrate their magnetic compass during migration. The research involved changing the magnetic field by 90° when the birds were resting at sunset. The imposed magnetic field was removed after sunset was completed, thereby leaving the birds in Earth's magnetic field. When the birds started off again (they were tracked by radio transmission), they altered their paths by 90° relative to a control group, which did not have the magnetic field direction reset at sunset. The authors' interpretation was that the thrushes were using the magnetic field for migration. They argued that since the magnetic field varies at different locations, the thrushes were repeatedly recalibrating the direction of true north relative to the local magnetic field direction by using the direction where the Sun set. Although this stimulating paper provides valuable information on the possible character of the magnetic map, it must be substantiated by additional studies: only a small number of individual birds were used in this study. In any case, the study also suggests there may be more than one mechanism by which birds sense Earth's magnetic field, consistent with the views emerging from an increasing number of experts in this area of research. Just as birds integrate substantial amounts of information to "see" colors, perhaps they are using two (or more) receptors—for example, a photochemical sense and a magnetite receptor—to sense different aspects of Earth's magnetic field and integrate this information with other input (such as the use of stars or polarized sunlight) to use in navigation.

*   *   *

A good candidate for the world's most ugly mammal is the naked mole rat, one of the many buck-toothed subterranean mammal species living in Africa. Mole rats, which use their long incisors to dig underground burrows similar to those of pocket gophers in North America, are more closely related to porcupines and guinea pigs than to either moles or rats. They have evolved to live in an atmosphere of low oxygen and high carbon dioxide, which characterizes their dark tunnel existence; their blood efficiently carries oxygen, and they have a low respiration rate. They are typically only 8 to 10 cm (3 to 4 inches) long and weigh about 30 grams (about 1 ounce). Mole rats are common throughout sub-Saharan Africa, where plants with long underground roots are common—the staple of mole rats' diet. All mole rats have rodent hair, except for the naked mole rats, which live in northern Africa. Some mole rats are able to move backward as fast as they can forward. Some also have the ability to use their large two front teeth separately, similar to the way humans use chopsticks.

While ugly in appearance, they are beautiful to study because they exhibit a wide range of social behavior. A few species of mole rats are eusocial, living in colonies with a hierarchy resembling that of insects, such as ants and termites. They have a queen, who is the largest and rules the colony with brute force; one or more kings; a soldier caste; and a worker caste made up of the smallest individuals. Naked mole rats (*Heterocephalus glaber*) of Ethiopia and Somalia sometimes have tunnels extending more than a kilometer in length, while more commonly mole rats have burrows around 100 meters long. The length of these tunnels is correlated with the number of individuals in a colony. The Zambian mole rat (*Cryptomys amatus*) has about 10 individuals per colony and digs the shortest tunnels, while the naked mole rats, which sometimes form colonies with more than 80 individuals, excavate the longest tunnels.[14]

The majority of mole rat species live a solitary existence, except when mating. One of these is the South African Cape mole rat (*Georychus capensis*), which has burrows averaging 130 meters in length. To find a potential mate, it thumps its hind feet on the burrow floor to produce seismic (Rayleigh) waves for communication.[15] In Namibia, the eusocial Damaraland mole rats (*Cryptomys damarensis*) are even claimed to find termites by using seismic waves.

Although I have often been in Africa, I have rarely seen mole rats because they spend their entire existence in dark tunnels about a meter or so below the surface. They come to the surface only when forced to do so. Mole rats only have slits for eyes and see poorly. The so-called blind mole rats even have a thin layer of skin covering their eyes and are essentially blind.[16] Yet all mole rats tested sense Earth's magnetic field, apparently to find their way through many branching tunnels, some of which house nests and food supplies. Clearly, this does not involve cryptochrome, a conclusion reached by all scientists studying mole rats.

In 1997 Stephan Marhold worked with Wiltschko and others to demonstrate that Zambian mole rats could sense Earth's magnetic field. Later the Israeli Tali Kimchi used an eight-arm maze to test Palestine blind mole rats' (*Spalax ehrenbergi*) navigational abilities. Although much larger, the maze looks something like a bicycle wheel, lying horizontally, with eight symmetric tunnels (spokes) extending outward from a central chamber to connect to a long circular tunnel (tire). Kimchi and others found that these blind mole rats prefer to locate their nests and food source in the southern part of this maze. When the magnetic field was shifted by 180°, the mole rats shifted their food storage and nests sites accordingly. Only the horizontal component of the magnetic field needs to be reversed (a 180° shift in the declination). The mole rats did not respond to changes in magnetic field inclination.

In 2006 German scientist Regina Wegner and her colleagues confirmed that Zambian mole rats preferred to make their nests in the southern part of a circular arena covered with opaque plastic that excluded all light. They then found small particles resembling magnetite in the mole rat corneas. To determine whether these particles were used for detecting Earth's magnetic field, Wegner and her colleagues anesthetized the corneas of many mole rats. Mole rats with anesthetized eyes could no longer sense Earth's magnetic field, while a control group treated with sodium chlorite solution could. Wegner and her colleagues concluded that mole rats are probably using magnetite particles in their corneas to sense Earth's magnetic field.

Bats are another group of mammals that can sense Earth's magnetic field. With an estimated 1,100 species, they constitute about 20 percent of the planet's mammal species. There are two major groups of bats, the fruit bats of the tropics, which do not echolocate, and the smaller new world bats, which do. Many bats migrate, sometimes thousands of kilometers, and even those that do not migrate often travel many kilometers from their roosts to feed at night or twilight. The majority of bat species use sonar, or echolocation, to prey on insects. They emit sound waves from their mouths or noses and use the reflections of those waves to sense the size and shape of, and distance to, objects. These sound waves have frequencies between 23 and 70 kHz (23,000 and 70,000 cycles per second), far higher than we can hear. Although this is an effective way of locating insects on which to dine, it does not allow them to find their way back to their roost, which can be many kilometers away.

In 2006 a multi-institutional group led by Richard Holland at Princeton University published the first magnetic field study of bats in the journal *Nature*. They used radio telemetry to track big brown bats (*Eptesicus fuscus*), which, like many bats, are not blind. They exposed one group of bats to a magnetic field rotated clockwise from magnetic north by 90° for 45 minutes before sunset and maintained the field for 90 minutes. They used the same protocol for a second group—only the field was rotated counterclockwise from magnetic north by 90°. The two groups of bats, which were about 20 km from their roost site, headed in significantly different directions, suggesting that the bats used Earth's magnetic field for navigation.

In 2008 Holland, Kirschvink, and their colleagues used the magnetic pulse method referred to earlier in this chapter to remagnetize single-domain grains in the big brown bat. When a pulse was delivered parallel to the direction of magnetization, the single-domain grains were not affected. As anticipated, the bats then flew in the correct direction to find their home roost. When the pulse was applied in the opposite direction, the single-domain magnetization inside the grain was reversed (appendix). It was anticipated that the bats would head off in the opposite direction from home. Although many did this, many others also flew in the correct homing direction. (The distribution was bimodal.) Holland and his colleagues concluded that bats use more than one mechanism to navigate and some override the signal from the single-domain grains to find the correct direction home.

In 2007 Chinese scientists Yinaw Wang and others found that bats (*Nyctalus plancyi*) preferred to roost in the magnetic southern side of a large cylindrical basket. When the declination was changed by 180°, the bats changed their roost to the opposite side of the basket. In contrast, they maintained their roosting position when only the inclination was changed. These bats only use the horizontal component of Earth's magnetic field. This is also apparently done by salmon, but not by turtles or birds.

Based on the limited number of experiments on bats, a good working hypothesis (used to refine experiments and theory) is that bats sense Earth's magnetic field and they use it, along with other sensory information, for navigational purposes. Experiments have also been conducted on other rodents not discussed here, but replication has sometimes proved difficult.

If some mammals can sense Earth's magnetic field, can humans also do so? One would think the answer is yes, based on the number of articles and blogs on the Internet, many of which cite Robin Baker's work in England. Joe Kirschvink, who once seemed enthusiastic about Baker's work, introduced me to Baker at a scientific meeting during the 1980s. Baker, then at the University of Manchester, described his experiments to me. After driving blindfolded students several miles on windy roads outside of Manchester, the students disembarked and correctly pointed to the direction of the starting point (within a few tens of degrees). In later experiments Baker followed the same experimental design, but half the students wore magnets on their heads while half wore dummy magnets. Baker claimed those students wearing dummy magnets could correctly identify the direction back to the starting point, while the ones with magnets could not. I asked him how the magnetic fields of other vehicles or power lines affected the students. His reply was something like, "Oh, do those matter?" I silently wondered how he had managed to get funding for his research. This evidently was a problem: Baker left the University of Manchester in the late 1980s, after failing to convince scientists reviewing his work that he should receive continual funding. In spite of this, Kirschvink thought Baker's studies hinted that humans might have a magnetic sense.

In 1992 Kirschvink and his colleagues found ferromagnetic material from extracts of human brain tissue and concluded it was most likely small uniformly magnetized magnetite grains.[17] Kirschvink then undertook more carefully designed experiments than conducted by Baker to see if humans could detect magnetic fields. These experiments were carried out in a specially designed room at Caltech, in which the magnetic field could be controlled. Kirschvink found no evidence that humans could sense magnetic fields. Other studies have confirmed this; it seems unlikely humans sense Earth's magnetic field. The magnetic particles in human brains appear to serve some purpose other than magnetoreception, or they are evolutionary remnants left over from our animal ancestors who used them.

Although humans don't appear to sense magnetic fields, electromagnetic fields can affect humans. Electrically charged atoms (ions) are involved in a host of processes in humans, including the conduction of nerve signals. A magnetic field (or an electrical field) will in most cases cause these ions to move. For example, transcranial magnetic stimulation (TMS) is an experimental technique approved by the U.S. Federal Drug Administration for research into reducing symptoms of such disabling diseases as epilepsy, migraine headaches, chronic pain, hallucinations, and obsessive-compulsive disorder. TMS involves the production of nerve signals in the brain triggered by rapidly changing strong magnetic fields applied to the head.

*   *   *

This is a good place to provide you with my summary of the animal navigation results. I have little doubt that many animals, ranging from bacteria to mammals, sense Earth's magnetic field, although humans appear not be among these. (Nevertheless, I find it puzzling that a broadly used sensory system seems to have been lost by humans. Perhaps aborigines in Australia or bushmen in Africa might be looked at to see if they have a magnetic "sixth sense.") I also conclude that some animals use the magnetic field for navigation. Many animals, such as birds, use other techniques as well, as evidenced by the fact that experienced animals (ones that have previously migrated) sometimes overrule magnetic information to find the "correct" navigation path. Curiously, for reasons not understood, some animals, such as turtles and birds, have evolved to use magnetic inclination, while others—including mole rats, bats, and salmon—have evolved to use the horizontal component of Earth's magnetic field.

The mechanisms employed by animals to sense Earth's magnetic field are varied. Magnetotactic bacteria almost certainly use their magnetic magnetosomes to sense Earth's magnetic field, whereas fruit flies and some birds require cryptochrome in their eyes to sense Earth's magnetic field. Other animals that can sense the magnetic field are blind and do not require cryptochrome. Some animals, such as birds, likely use more than one magnetic sensory device. It makes sense that they might coordinate the use of cryptochrome, which is also used to determine the circadian rhythm to tell them when to migrate, with some other magnetic sensor, such as single-domain magnetite particles, to solve what we humans call the map problem. I believe many animals also employ other non-magnetic senses to solve the map problem: for example, the recall of topographical features, the polarization of the Sun, and the use of stars. However, there is no animal that we can point to and say, "We know how this animal has solved the map problem."

CHAPTER SIX

The Effects of Geomagnetism
and Plate Tectonics on Climate
and Paleoclimate

The mountain environment has changed over the few decades I have been hiking and climbing. The glaciers are generally receding, and the animal and plant life is adapting to the hydrological changes this has brought. Does this reflect that we are still emerging from an ice age? Have variations in Earth's magnetic field affected our climate? Can the magnetic record stored in rocks provide any answers to these questions? We ponder such questions in this chapter.

Weather and climate are different. Climate refers to the state of the atmosphere and its variability, characterized by properties such as temperature and precipitation, averaged over several decades or longer. In contrast, it is a weather forecast, rather than a climate forecast, you want when you are planning a picnic the next day. In brief, climate is the average of weather over a long time interval.

I am often asked how climate has changed in the distant past (paleoclimate). How do scientists sort out natural changes in climate from man-made ones? In spite of considerably large changes in the factors causing climate changes, our climate has varied less than might be anticipated. The change in temperature of any object depends on how much heat is supplied to it relative to how much heat it loses. If less heat is supplied than is lost, the object's temperature decreases and vice versa. As the object's temperature decreases, it loses less heat to the surroundings. Eventually, at least in theory, equilibrium is achieved when the object reaches a state where the amount of heat lost is precisely balanced by the amount of heat gained. However, Earth and its atmosphere are complex, and at most only a quasi-equilibrium state is ever achieved. Earth has been slowly cooling from its initial hot state. But when Earth formed, the Sun was 30 percent less bright than it is today, an estimate based on theoretical calculations and on the observation of similar stars in our galaxy in different stages of their evolution. Our earliest atmosphere, which formed primarily from gases released in volcanic eruptions (meteorites also contributed), was rich in carbon dioxide and devoid of oxygen. Substantial amounts of oxygen entered our atmosphere between 2.5 and 2 billion years ago in what is known as the great oxidation event. This event corresponded with the expansion of cyanobacteria, bacteria that obtain their energy by photosynthesis, which involves the conversion of carbon dioxide to carbohydrates and oxygen.[1] Concomitant with the early increases in oxygen were decreases in carbon dioxide and methane, greenhouse gases that kept Earth's lowermost atmosphere above freezing temperatures during times when the Sun's radiation was substantially less than today's. Through all of these changes, Earth's atmospheric temperatures have varied little. Oceans have existed for at least 3.7 billion years without being completely frozen or without evaporating away. This is fortunate because liquid water is essential for life as we know it. The magnetic field of Earth has also played an important role, as our atmosphere and its chemical composition would be substantially different if the magnetosphere did not offer substantial protection from the Sun's solar wind (chap. 4).

No individual is an expert on every factor affecting climate. Scientists enter information on radiation, greenhouse gases, clouds, ocean-atmospheric interactions, and so on into mathematically based computer programs to gain insight into the processes giving rise to the present and future climates. A global circulation model (commonly called a global climate model), or GCM, is used to forecast climate for the entire Earth. GCMs, the most detailed computer simulations available for predicting climate change, evolve as our knowledge grows. Because different GCMs use different input data, different assumptions, and different mathematical techniques,[2] their forecasts also differ. Most GCM predictions of global warming in this century range between 1°C and 6°C.

You might wonder how GCMs are checked. This is done in different ways, including by hind casting: modelers input parameters from the past to determine whether they correctly predict past climates. Often they don't in all respects. Then the model is adjusted. Modelers might change the parameters affecting, say, cloud or sea ice evolution until they obtain reasonable agreement with observations. Many different GCMs reasonably "predict," in retrospect, the climates we have had during the past century. However, all make somewhat different assumptions, which lead to different future outcomes. This provides some room for skepticism. Indeed, when one reads various blogs on the Internet, one becomes aware of how nasty some aspects of the debate about climate forecasting have become. Perhaps the nastiest is how magnetic field variations impact global warming, a subject I return to in the next section.

Usually the public is presented with oversimplified interpretations of what causes global warming. Consider atmospheric greenhouse gases, of which there are many, including water vapor, carbon dioxide, methane, nitrous oxide, and sulfur dioxide. The absorption and re-radiation by greenhouse gases slows the rate our planet loses heat. This causes our atmosphere to heat up. The media emphasizes one of these, carbon dioxide, which humans are adding to our atmosphere by burning fossil fuels and through deforestation. Carbon dioxide is estimated to account for about 10 to 25 percent of Earth's greenhouse effect in 2008. (Water vapor is estimated to account for 36 to 70 percent of the greenhouse effect.)

As seen in figure 6.1, changes in the temperatures in our atmosphere have been closely correlated to changes in carbon dioxide for hundreds of thousands of years—long before humans had any measurable input. When the temperature is higher, so is the carbon dioxide. This strong correlation has now been extended to 800,000 years before present in one Antarctic ice core and for millions of years using oceanic sediment core data. Carbon dioxide has risen from around 280 parts per million per volume (ppmv) in the 1850s to 383 ppmv in 2008, the highest level in the last several hundred thousand years. It is sometimes argued that the correlation between carbon dioxide and temperature demonstrates beyond all reasonable doubt that increasing carbon dioxide causes temperature to increase. But the science behind this correlation is more complicated.

Many factors affect climate, including the Milankovitch cycles (chap. 1). The approximately 100,000-year variation in the temperature and carbon dioxide shown in figure 6.1 is often attributed to the Milankovitch cycle dealing with the eccentricity of Earth's orbit. (There are also other hypotheses. The variation in dust shown in figure 6.1 is discussed later.) Earth orbits the Sun in an ellipse, whose shape changes with time. For example, sometimes its length (long axis) is smaller, which, on average,

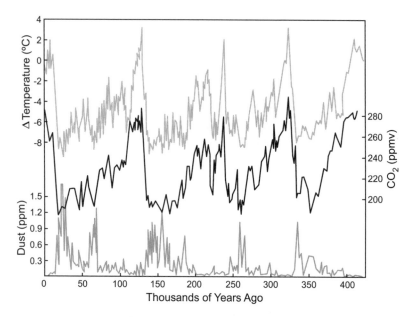

FIGURE 6.1   Data from the Vostok (Antarctica) ice core are shown as a function of time. Carbon dioxide (middle curve), reconstructed temperature (top curve), and dust (bottom curve) for the past 420,000 years. The temperature is determined using procedures such as discussed in note 14 in chapter 1. The 100,000-year cycle is most easily seen in the carbon dioxide and temperature curves. Figure modified by Beth Tully from *Wikimedia Commons*.

brings Earth closer to the Sun, while at other times it is greater. This produces a small 100,000-year variation in the amount of heat our planet receives from the Sun, and possibly explains the temperature variation shown in figure 6.1.[3]

But why does carbon dioxide show this same periodicity? Atmospheric scientists conclude that various feedbacks must be involved, such as the one involving the solubility of carbon dioxide in the oceans and temperature. As the atmosphere's temperature rises, the oceans also become warmer. Because the solubility of carbon dioxide in water decreases as temperature increases, warmer oceans absorb less carbon dioxide from the atmosphere. Therefore, the atmosphere's carbon dioxide increases, as does its temperature through the greenhouse effect. This feedback between the oceans and the atmosphere continues until quasi-equilibrium is reached in which both the atmosphere and oceans are warmer. A quantitative estimate using computers is required to determine whether this feedback is sufficient to explain the correlation shown in figure 6.1. According to experts, calculations indicate that some, but not the preponderance, of the correlation between temperature and carbon dioxide shown in figure 6.1 is explained by the feedback involving carbon dioxide solubility in the oceans. Other feedbacks must be involved. Perhaps they are associated with changes in the amount of photosynthesizing plants, or ocean circulation. No one really seems to know for sure. There are many other uncertainties in accurately predicting changes in Earth's climate, which lead to conflicting interpretations. How can a non-expert evaluate these?

I am not an expert in atmospheric science and climate change, even though I have ready access to atmospheric science faculty, who occupy the offices on the floors immediately above mine at the University of Washington. How, then, does someone like me decide what to believe? I do the same thing many of you do: I rely on an expert's opinion. This is not valid in logical terms. Authority arguments, such as "Einstein said this was true of relativity, and therefore we should believe it" are invalid in a strict logical sense. Even Einstein made mistakes while developing his theories of relativity, and parts of the theory have been extended beyond his original contributions. We accept the view of authorities because we believe they have a higher probability of being correct than non-experts. Because of such reasoning, scientists—like almost everyone else—commonly use authority arguments. For example, many scientists accept authority arguments when consulting with a medical doctor or accept reasons for a car malfunction from a car mechanic. In many instances an authority argument, although not logical in a formal sense, is a reasonable choice because of the time it would take to access the situation from first principles. However, we should not fool ourselves when we do this. We don't know the correct answer, but we are assuming our expert does. This means we should choose our experts with considerable care. In particular, we should check what they tell us by using things we already understand. In other words, we should calibrate our "experts." The more important the subject is to us, the more work we must do to understand various aspects of it to evaluate experts.

It makes no sense for me to tell you what to believe on global warming, as I am not an authority. I can tell you the science is typically far more complex than the popular media implies. Large uncertainties in our ability to accurately forecast climate change remain. One relatively conservative authority is the U.S. National Academy of Sciences. Although different predictions are made by different GCMs, the National Academy of Sciences reported that all reliable models predict a significant increase in temperature this century, even if we immediately cut back on the burning of fossil fuels. Naturally, the amount of global warming will be larger if we don't do this. The National Academy also concluded that both anthropogenic and natural processes were significant in having led to an increase in the average surface air temperature during the past century estimated to be between 0.4°C and 0.8°C (0.7°F and 1.5°F). The warming trend is consistent with the global retreat of glaciers, the reduction in sea ice, and so on. Even some animals, such as birds, might have already altered their times of migration and breeding. Climatic change is affecting life on a global scale. The GCMs predict a global warming of 1°C to 6°C by the end of the twenty-first century. Other large organizations, such as the Intergovernmental Panel on Climate Change and the American Geophysical Union, arrive at similar conclusions.

*   *   *

"There is no convincing scientific evidence that human release of carbon dioxide, methane, or other greenhouse gases is causing or will, in the foreseeable future, cause catastrophic heating of the Earth's atmosphere and disruption of the Earth's climate."

So began the first sentence of the second paragraph in a petition I was requested to sign. The petition began with, "We urge the United States government to reject the global warming agreement that was written in Kyoto, Japan, 1997, and any other similar proposals." It was part of a thick envelope I received in 2007, which also contained a copy of an outdated *Wall Street Journal* newspaper article, a recent scientific paper, and a letter from Frederick Seitz, past president of the National Academy of Sciences and president emeritus of Rockefeller University. Seitz had been one of the pioneers of solid-state physics during the first half of the previous century. He went on to become the president of the National Academy of Sciences in 1962. The scientific paper included in the envelope was published in the *Journal of American Physicians and Surgeons* in 2007, a curious place to publish a paper on global warming. Seitz had sent out a similar petition in 1998, which was signed by thousands of American scientists. I did not sign either petition. When he died in 2008 at the age of 96, Seitz was at odds with the prestigious body over which he once presided, the National Academy of Sciences, who warned the public that humans were contributing to global warming. I expect various groups will use Seitz's petitions to argue that it remains controversial among scientists whether the burning of fossil fuels significantly increases global warming, causing adverse effects on our environment. In contrast, a strong majority of knowledgeable scientists attribute a significant amount of the recent global warming to humans.

Most scientists who signed Seitz's petitions, including many with only an undergraduate degree in an unspecified field of science, would not have been reasonably able to evaluate the scientific paper included in the envelope from Seitz claiming that the "computer models are markedly unreliable" and denying that human input into global warming has any significant effect.[4] The paper claims hydrocarbon use is not correlated with temperature increase during the previous century, but solar activity is. A few geomagnetists also think solar activity is a prime culprit, such as Vincent Courtillot, a prestigious geomagnetist who was deputy to Claude Allègre in the French government. Claude Allègre, a talented but sometimes abrasive geochemist, was the French minister for education from 1997 to 2000. Allègre describes Al Gore as a "crook" in a 2007 book, *Ma vérité sur la planète* (My truth about the planet), for promoting falsehoods on global warming to foster select business interests.

Over the years, I have had many discussions with Courtillot, presently the director of the Institut de Physique du Globe de Paris. I find him to be a very bright individual who loves a good debate, which is often one-sided because he talks quickly, is knowledgeable, and has strong emotional attachments to his arguments. I recall one three-hour lecture I received at dinner after I mentioned I agreed with most earth scientists that a bolide (asteroid or comet) collided with Earth 65 million years ago to cause the extinction of dinosaurs and many other species (the Cretaceous-Tertiary, or K-T, extinction). More than 50 percent of all the then-existing species are estimated to have gone extinct at this time. Courtillot thought the K-T extinction was due to a vast outpouring of basalts in India referred to as the Deccan Traps. He believed that these massive floodplain basalt eruptions injected enough material into the atmosphere to alter our climate and eliminate a majority of species living at that time. I had made my comment not knowing he was writing a book on the subject. He was quick to point out another major extinction at 252 million years ago that was associated with a massive outpouring of basalts in Siberia. It is the largest of the five major extinctions recognized in the geological record. Some paleontologists estimate that more than 90 percent of all known species went extinct 252 million years ago. Although speculations abound, no consensus exists as to what caused this granddaddy of extinctions.

There is a consensus that an asteroid was responsible for the K-T extinction 65 million years ago. However, science is not done by consensus; sometimes the minority point of view prevails. A minority of scientists thinks the eruption of the Deccan Traps' basalts affected our climate sufficiently to put many species in a precarious position. When an asteroid hit the Yucatán (Mexico) peninsula 65 million years ago, it merely delivered the final blow to many animals and plants. Others think the asteroid played a negligible role. The majority of geologists argue against these views, pointing out that many species died off abruptly 65 million years ago, rather than having been spread out over the million-year time span during which most of the Deccan Traps' lavas erupted. Courtillot seems to enjoy controversy, and he always has thought-provoking comments to make on practically any subject under discussion. In the past few years, Courtillot and his colleagues have taken on another unpopular challenge: to convince scientists that variations in solar activity manifested in sunspot cycle lengths and intensity contribute significantly to the recent global warming. A consequence of this, they argue, is that the human input is small or negligible. I will return to such arguments momentarily.

The history of attributing climate change to sunspots goes back at least to the beginning of the nineteenth century when in 1801 the astronomer William Herschel (1738–1822) suggested that an increase in the number of sunspots was correlated with a lower wheat price. Later the economist William Jevons (1835–1882) suggested that business cycles might be affected by variations in the number of sunspots. He presented a statistical analysis linking commercial crises to the 11-year solar cycle. He argued the 11-year sunspot cycle affected weather, which in turn affected corn production. This appears to be the first use of time-series analyses to relate the 11-year solar cycle to any other phenomenon.

A time series refers to a sequence of numbers taken over time; for example, a time series of the average annual temperature for each year during the twentieth century would consist of a series of 100 numbers. Mathematical techniques such as Fourier analyses (chap. 1) are used to determine the properties of time series. Such techniques, which have multiplied during the past century, are used to determine trends and periodicities in a time series. They have also led to data mining, in which scientists attempt to discover subtle signals buried in noisy (messy) data. Many arguments over interpretations of geomagnetic effects on climate involve the alleged use of inappropriate data selection or improper statistical analyses. "Inappropriate data selection" often refers to cherry-picking: scientists use data supporting their beliefs while ignoring data that don't.

Data-mining statistical techniques don't always yield the correct answers. Consider the numbers 1, 3, −3, 4, 4, −7, 4, −1, −2, 2, 3. Suppose each number represents the deviation in inches of the total rainfall in some city over ten years, compared to the average rainfall during the previous century. Therefore, the total series represents slightly more than a century's worth of data. Every third number in this series is negative. All other numbers are positive, except the eighth number, which is slightly less than zero. A graph of this series indicates four relatively evenly spaced maxima separated by three minima. The use of many modern time-series analyses could lead one to conclude that periodicity is present in this series: less rain falls every third decade (a minus number) compared either to the previous century or to the preceding decade. But we should not believe this conclusion. These numbers were generated from the number pi, which, as an irrational number, cannot exhibit any real periodicity.[5] The apparent periodicity occurred because the time series was not long enough. The numbers following the ones given above are 1, −2, −6, −1, 1, 5, −4, 2. The alleged periodicity is not sustained. Similarly, Jevon's conclusion that variations in the 11-year sunspot cycle were correlated with variations in crop productivity broke down when more data became available to provide a longer time series. Periodicities are often claimed, when none exist, because of insufficient data in a time series.

Another common error occurs when a priori information is available. Suppose I ask a scientist to calculate the probability a car with license number DMN 751 is in a certain parking lot. Assuming every letter (there are 26 possibilities) and number (10 possibilities) are randomly selected, the probability appears to be 1 in 26 × 26 × 26 × 10 × 10 × 10 = 17,576,000. The probability is less than 1 in 17 million. Although this is a very low probability, I may know the car is there because it is mine. I have a priori information concerning the car that was not provided to the scientist. Another example of a priori information is the puzzle in which a quiz-show host rewards a contestant with a prize lying behind one of three doors. A goat lies behind two of the doors and a car the contestant wants is behind the third door. The contestant chooses one of the doors. The host, who knows what is behind each door, opens one of the doors not selected by the contestant to reveal a goat. He asks the contestant if he would like to change his choice. Should the contestant do this to optimize the chances of getting the car? Although the answer is given in note 6, see if you can calculate the probability.[6] I have provided you with a big hint: the host has a priori information, which affects the probability analysis. These examples will prove useful to us in analyzing possible geomagnetic effects on weather.

Some of my geomagnetic colleagues claim it is clear that solar magnetic fields affect climate. They say that the Maunder minimum, the interval of time between 1645 and 1715 with almost no sunspots (fig. 4.2), caused the Little Ice Age. Although many scientists agree with this conclusion, are they correct? The Little Ice Age is best documented in Europe and North America. It was a time in which it appears glaciers on average expanded rather than contracted. The Little Ice Age had an uncertain beginning, but ended around 1850. Nevertheless, most scientists acknowledge the cooling in the Little Ice Age was most pronounced during the Maunder minimum. Advocates posited that a decrease in solar activity led to the cooling of our atmosphere by processes discussed later.

One competing hypothesis is that the Little Ice Age was caused by heightened volcanic activity. Large explosive volcanic eruptions often eject ash and gas, including sulfur dioxide, into the stratosphere, which extends upward from the top of the troposphere to about 50 km. The sulfur dioxide interacts with water vapor (also ejected by these volcanoes) to form sulfuric acid, which rapidly condenses to form submicroscopic droplets called aerosols. Another volcano-produced aerosol is ash. Aerosols have two opposing effects on temperature: they can absorb light to heat the atmosphere, or they can increase Earth's albedo (more solar radiation is reflected back into space) to cool Earth. In the case of volcanoes, the sum of these effects is usually thought to produce cooling. For example, Mount Pinatubo erupted in the Philippines in June 1991, sending sulfur dioxide and ash more than 20 km into the stratosphere. While the precise effects of this eruption on our troposphere are not fully understood, most atmospheric scientists conclude that the eruption of Mount Pinatubo caused a slight global cooling lasting perhaps up to three years after the eruption. Some earth scientists suggest that large eruptions were more common during the Little Ice Age and caused cooling on a global scale.

Yet another hypothesis suggests that a variety of factors affecting climate just happen to have coincided to produce cooling during the Little Ice Age. Many of these factors involve the ways in which the atmosphere and oceans are coupled. El Niño ("little boy" in Spanish) was identified by oceanographers as a condition in which the trade winds waned to allow warm waters to build up in the eastern tropical Pacific Ocean. (La

Niña, or "little girl," approximately represents the opposite conditions.) The buildup of warm water impedes the upwelling of nutrient-rich waters off the coast of Peru and has a marked effect on fisheries. For much of the twentieth century, El Niño was thought to be a relatively local phenomenon. With time, scientists found that El Niño conditions affected climate on a global scale. For example, Australia was warmer and drier, and southern South America was wetter than usual, during an El Niño event. Changes in oceanic currents have profound effects on climate. At the same time oceanographers were studying the effects of El Niño, atmospheric scientists were studying a phenomenon called the Southern Oscillation. In the 1920s Sir Gilbert Walker (1868–1958) noticed that high-pressure conditions in the Pacific often corresponded to low-pressure ones in the Indian Ocean. It eventually became clear that El Niño and the Southern Oscillation are associated phenomena relating complicated feedbacks between the atmosphere and the oceans. Scientists now refer to this as ENSO (El Niño–Southern Oscillation).

Intervals of time between major ENSO events are irregular, ranging from 2 to 10 years, with each ENSO event lasting on average about 2 years. ENSO reflects but one of several complicated interactions between the oceans and the atmosphere. Another is the Pacific Decadal Oscillation (PDO), which describes a shift from a warm phase of the western Pacific Ocean north of 20°N to a cool phase. During a warm phase, the western surface waters are warmer and the eastern waters are cooler. Cool phases, which reflect opposite conditions, occurred from 1890 to 1924 and from 1947 to 1976. We may also have entered a new cooling phase near the beginning of this century. As you can see, the "decadal" cool phase refers to a 20-to-30-year interval.

Although it is debatable whether any true periodicities occur in ENSO, PDO, or other oscillations (not discussed), the way heat is stored in the oceans and carted around by currents clearly has a strong effect on our atmosphere. In turn, changes in atmospheric winds, such as the waning of the trade winds in the Pacific Ocean, can have a strong influence on oceanic currents. The complicated (non-linear) manner in which the atmosphere and oceans are coupled can give rise to climate changes acting on a wide range of time scales. When some of these changes reinforce each other, an extended interval of cooling such as the Little Ice Age might occur.

Scientists need to devise tests to sort out competing hypotheses purported to produce the Little Ice Age. Although some advocates claim that the accidental coincidence of the Little Ice Age with the Maunder minimum is too improbable to believe and further tests are not needed, they did not consider that a priori information was available. It is an improper statistical analysis to calculate that the probability the Little Ice Age and the Maunder minimum coincide after it has already been established that they do. Such claims are similar to those associated with the probability calculated earlier of finding a particular car's license in a parking lot and should not be believed. However, it is legitimate to form a hypothesis based on the coincidence of the Maunder minimum and the Little Ice Age. A fair test of the hypothesis is to examine other times in the geological record when solar activity was low to determine whether they coincided with times when Earth's temperatures were also lower. Ice core data indicate periods of cooling near 1400 years before present (BP), 2400 years BP, 4200 years BP, 5900 years BP, 8100 years BP, and 9400 years BP. These can be compared to intervals of lower solar activity that have been identified using chemical isotopes.[7] The isotope data do not indicate a significant coincidence of inferred solar minima with intervals of significant cooling. Nevertheless, advocates claim that the presence of 90- and 200-year periodicities in the solar cycle, based on analyses of isotope data, can be combined in complex ways (including time lags) to establish a connection. I find such arguments unconvincing: even the 90- and 200-year claimed periodicities have not been well established by robust time-series analyses. Although many geomagnetists accept that the low solar activity during the Maunder minimum caused the Little Ice Age, I consider it an interesting hypothesis that presently has only weak support.

Although many correlations linking the 11-year solar cycle to various climatic data were claimed, by the middle of the twentieth century such correlations were widely considered to be pseudoscience. Therefore, many scientists were surprised when data near the end of the twentieth century demonstrated a clear link between the solar cycle and climate. Starting with the Solar Maximum Mission in 1980 and confirmed by later satellite missions, convincing evidence emerged that climate is affected, at least in a minor way, by the 11-year solar cycle. The total radiation our atmosphere receives from the Sun varies by about 0.1 percent over the 11-year cycle: it is highest at solar maximum and lowest at solar minimum.[8] In this case, the minority point of view trumped the majority one claiming that the solar cycle had no effect. This provided ammunition to scientists who support the speculation that magnetic activity significantly affects our climate. It also seems likely that a 0.1 to 0.6 percent variation in the total solar radiation occurred since the Maunder minimum. These changes are associated with variations in the length and strength of individual sunspot cycles, as shown by the dark line in figure 4.2.

The direct effect of a 0.3 percent variation in solar radiation has been shown to have a minor effect on our climate by using GCMs, the computer models of climate. (The 0.3 percent value was used because it was near the middle of estimates for the change in solar radiation since the Maunder minimum.) Could an indirect process amplify some of the temperature variation? This is an important question because the small variations in heat associated with the Milankovitch cycles, such as those reflected in the 100,000-year temperature cycle shown in figure 6.1, are too small by themselves to have much effect on the atmosphere's average annual temperatures. Many scientists think that the temperature must be amplified by complex (nonlinear) processes (see note 3). Perhaps, as some scientists hypothesize, feedback processes also amplify the effects of solar radiation.

Satellite measurements indicate that the variations in the solar radiation are not uniformly distributed over all wavelengths. In particular, the ultraviolet radiation, the kind that produces sunburn, is particularly large. The absorption of ultraviolet light by ozone heats the stratosphere. A feedback exists in this system because ultraviolet light also enhances chemical reactions in the stratosphere, referred to as photochemical reactions, which produce ozone (containing 3 oxygen atoms per molecule) from oxygen (containing 2 oxygen atoms). As the ozone increases, so does the stratosphere's temperature. In this case, however, the feedback is not large: the temperature is only increased by around 0.02 percent in our stratosphere—not enough to have a significant effect on the temperature at Earth's surface. Geomagnetists have found other mechanisms that might have a much larger effect, as we will discuss after examining evidence suggesting that variations in solar output might affect our climate.

Time-series analyses of a variety of data suggest to some scientists that periodicities in Earth's magnetic field might be linked to variations in climate. Other scientists doubt the fidelity of the record or the reliability of the time-series analyses. No consensus exists on whether any of these periodicities are reliable and, if they are, what they imply.[9] Often the claimed links between solar magnetic field variations and climate used indirect evidence, as illustrated by the provocative studies of Australian geologist George Williams.

While I was in Australia in the 1980s, I met Williams, a geologist who used data from approximately 700-million-year-old sediments to convince many of my scientific friends that solar activity had a pronounced effect on our climate. The sedimentary rock formation in southern Australia that Williams studied exhibits 19,000 alternating layers of sandstone and siltstone. The layers are remarkably evenly spaced and, on average, every twelfth layer is more pronounced: it is darker and thicker. Sometimes as few as 8 intervening layers separate successive pronounced layers and some-times as many as 16 intervening layers do. Williams thought the layering reflected changes in the amount of sediments delivered by runoff into a large lake over the course of a year. The sands were deposited by faster-flowing water during spring and summer, while the silts were deposited about 6 months later when the flow was less. Although Williams had no absolute age determinations, the annually deposited layers could be used to estimate that the entire formation was deposited in about 19,000 years. The variation in the layering was remarkably similar to the variations in the number of years in different solar cycles, which typically vary in length from 8 to 13 years (chap. 4). Williams hypothesized that this implied a direct causal relationship. In a paper in 1981, he advanced the hypothesis that the 11-year solar cycle 700 million years ago was remarkably similar to the present one and that it significantly affected Earth's climate in the past and continues to do so today.

Over the next few years, I declined separate requests by an Australian geomagnetist and an American physicist to work on theoretical mecha-nisms by which the solar cycle affected Earth's climate. Both requests had come from scientists aware of Williams's paper. In 1988 Williams reinter-preted the layering as reflecting tidal rhythmites: the layers were deposi-tion during a time of large tidal variations in an estuary. The alternating layers mostly reflected daily variations in tidal flows. Stronger tidal currents carried more sand, and lesser ones carried more silt. The new interpreta-tion implies the entire formation had been deposited in slightly more than a half of a century, rather than 19,000 years. The connection to the solar cycle vanishes in this interpretation, which is presently favored.

Let's consider a few of the reasons why some geomagnetists, includ-ing Courtillot, think there is a significant link between solar activity and climate. As noted in chapter 4, geomagnetists use a wide variety of indi-ces to describe magnetic activity associated with the external magnetic field as recorded by magnetometers at Earth's surface. One of these is the so-called official aa index, which uses measurements to characterize short-term variations in the external magnetic field made at two nearly antipodal observatories in Australia and England. Although there is a wide use of the aa index to characterize variations in the solar magnetic field, some geomagnetists prefer to use an aa index obtained from other observatory data. For example, a much debated study published in 2007 by Courtillot and his colleagues uses data from magnetic observatories in Scotland and Alaska, which, being at higher latitudes, are argued to give more reliable information on magnetic storm activity (chap. 4). Although the details of data selection and related issues are important to specialists, I will greatly oversimplify this discussion by considering average sunspot activity as indicated by the smooth black curve in figure 4.2. In particular, solar activity increased from the beginning to about the middle of the twentieth century, after which it leveled off and decreased for about 20 years. This coincides with a decrease in global temperatures, a decrease that occurred even though carbon dioxide in the atmosphere was increasing.[10] Advocates conclude that solar magnetic field effects on climate are large enough to overwhelm the effects of greenhouse gases. Not only can variations in the solar wind explain the Little Ice Age; they can explain the gradual warming thereafter and even some of the detailed changes in the twentieth century.

Most atmospheric scientists disagree. This was reflected in the reports of the National Academy of Sciences, the Intergovernmental Panel on Climate Change, and the American Geophysical Union, organizations that considered such information in arriving at their conclusions given at the beginning of this chapter. Their conclusions are supported by evidence that an increase in global temperature occurred after 1980, which is attributed to a man-made increase in carbon dioxide, while solar activity decreased (see fig. 4.2). Atmospheric scientists also say they have explanations for the decline in temperatures beginning near the middle of the previous century that do not involve solar activity. Some claim that the cooling comes from increased aerosol pollutants, while others attribute it to natural variability in our climate. Such arguments reflect the complexity of processes affecting our climate. This is further illustrated by the suggestion that we may experience a decade or so of cooling, before greenhouse gas effects again drive global temperatures upward. This is based on recent (2008) evidence of a slowing of heat-carrying Atlantic Ocean currents, which leads to a cooling of western Europe, North America, and a slight overall global cooling. Such complexities illustrate the necessity of using as long a time series as possible in attempting to sort out climate fluctuations. Also because of such complexities, magnetic field advocates must convince skeptics that a viable mechanism linking solar variation to climate exists.

Although a few speculations emerged to explain possible correlations between solar activity and climate, the one dominating the present discourse emerged in 1997 when Danish scientists Henrik Svensmark and Eigel Friis-Christensen suggested that the link between solar activity and climate involves galactic cosmic rays. (This was updated a year later by Svensmark.) "Cosmic rays" is a misnomer. No rays are involved. Instead 90 percent of cosmic rays are just energetic protons that come from the Sun or space. Most of the rest of the "rays" are helium nuclei. Galactic cosmic rays have higher energies (higher velocities) than solar ones. They impact Earth from all directions, and most are thought to originate in supernovas. Svensmark and Friis-Christensen recognized that cosmic rays were identified in a cloud chamber (supercooled and saturated with water or alcohol vapor) by a cloud streak. In 1997 they claimed that total cloud cover was greater from 1984 to 1991 due to a greater influx of galactic cosmic rays. In a paper in 2000, Nigel Marsh and Svensmark further developed this argument by presenting evidence linking low solar activity to an increased abundance of clouds lying below 3 km. When solar activity is most intense, it produces, on average, a stronger solar wind. This deflects the galactic cosmic rays more than when the solar wind is weaker. Hence, more galactic cosmic rays enter our atmosphere when solar activity is relatively low, perhaps, as Svensmark and his colleagues speculate, allowing more low-altitude clouds to form. The mechanism, they say, is that the galactic cosmic rays cause ionization in our atmosphere, which allows clouds to nucleate easier.

Water droplets in clouds form around nuclei, submicroscopic particulate matter referred to as aerosols. Aerosols come from many sources ranging from volcanic eruptions to biological ones, even bacteria. For example, sulfate aerosols can be produced in the atmosphere from volcanoes, which emit sulfur dioxide, from fossil fuel burning yielding sulfur dioxide, and from marine phytoplankton algae that emit dimethyl sulfide. Because ions are too small to be effective by themselves to nucleate a water droplet, Svensmark and his colleagues suggest they act as catalysts to form effective aerosols. However, they do not clearly specify how this works. Atmospheric scientists sometimes refer to this as mysterious physics. Even if ionization leads to aerosol production, atmospheric scientists are skeptical that such a source could effectively compete with other known aerosol sources.

Clouds are one of the most difficult factors to treat in any model for climate. The manner in which they are modeled produces large differences in GCM predictions on global warming. One important way that clouds affect climate is through the scattering of solar radiation. The oceans are the darkest regions on our planet: they have an albedo, the proportion of reflected light, ranging from a low near 6 percent at low latitudes to values about three times this at high latitudes.[11] In contrast, clean snow reflects almost all of the incoming radiation. The albedo of clouds varies from about 10 to 90 percent, depending on circumstances such as cloud thickness, water-drop size, and ice content. On average, about two-thirds of our planet is covered by clouds at any given time. Their distribution, evolution, and structure affect climate in complex ways. For example, clouds produce a blanketing effect, which heats our planet. The degree of blanketing depends on many factors, including the type of cloud and the size and distribution of its water droplets and ice crystals. In contrast, solar radiation is reflected from the top of clouds to cool our planet.

Although complex, atmospheric scientists conclude that the average effect of clouds is to cool Earth's atmosphere. Low-lying stratus clouds above the oceans are particularly effective at doing this. These are the very clouds that Marsh and Svensmark claimed are correlated with low solar activity. Nevertheless, in 2003 Peter Laut published a paper claiming that the time series used by Marsh and Svensmark was too short and that calculation errors occurred during the processing of the data. This led to more claims and counterclaims in the scientific literature. Regardless of the details, the correlation alleged by Marsh and Svensmark has not been confirmed by more recent data. For that matter, all time series used by critics and advocates alike appear too short to me: adequate satellite data have only been available for the past three decades, and climate processes are complex. I expect to see many more claims and counterclaims before this matter is settled.

Svensmark and his colleagues have recently carried out experiments in the laboratory to demonstrate a proof of concept: clouds can be produced under laboratory conditions when cosmic rays pass through atmospheric gas containing water vapor and sulfur dioxide. Because of doubts about how well the laboratory conditions represent natural ones, these experiments have not convinced many atmospheric scientists. However, they appear to have convinced others. In 2007 Svensmark and his colleague Nigel Calder published *The Chilling Stars: The New Theory of Climate Change*, in which they claim that galactic cosmic rays affect climate more than man-made carbon dioxide. The evidence presented by Svensmark and his colleagues was sufficient to persuade administrators and scientists at CERN, the European Organization for Nuclear Research, in Geneva to initiate a project using an accelerator to produce protons and other particles with sufficient energies to mimic galactic cosmic rays to study the cloud nucleation process. Results of those experiments should be forthcoming within a few years.

Let's summarize the present situation. Most experts agree with the summaries of prestigious organizations such as the National Academy of Sciences, Intergovernmental Panel on Climate Change, and the American Geophysical Union that global warming is occurring and is primarily driven by an increase in greenhouse gases, which man has significantly contributed to in the past century and continues to do so today. Experts point out that it is relatively easy for scientists to estimate the temperature rise associated with an increase in carbon dioxide. For example, advocates argue that a doubling of carbon dioxide from human input over its value in 1850 would lead to an increase in 4 watts per meter squared over the entire surface of Earth. This is equivalent to the energy emitted from a 100-watt lightbulb in a 25-meter-squared room. Although this may seem like only a small amount of additional heat, remember this is for every 25 square meters of Earth's surface. If the temperature is not increasing because of this, some other (unknown) effect must be occurring to counter it. Although confounding effects are undeniably present, at least on a scale of a few decades, the average upward trend in temperatures will continue throughout this century because of greenhouse gases. These prestigious organizations also conclude that variations in solar activity can affect climate, at least in a minor way, as evidenced by slight changes in the total radiation received from the Sun. But, they conclude, the climate effects caused by variations in solar activity do not change the big picture showing that greenhouse gases warm our climate.

The majority of experts could be wrong! Some advocates even claim that variations in solar activity have a more pronounced effect on climate than does the burning of fossil fuels and deforestation. Supporters on both sides of the impact of solar variation on climate often openly complain that opponents cherry-pick their data, use time series that are too short, and apply inappropriate data-mining types of statistical analyses. In private (or usually so), opponents accuse each other of putting self-interest ahead of scientific integrity. Some even accuse opponents of scientific fraud and suggest they should be jailed. Putting aside the rhetoric, the advocates supporting a major role for solar variability affecting climate will not be widely believed unless they produce a viable mechanism. For example, there must be a clear demonstration that galactic cosmic rays produce clouds under natural conditions. This mechanism must also be quantified sufficiently to be used in computer models to make testable predictions. Unless those who support solar magnetic field forcing can do this, I suspect they will remain in the minority.

*   *   *

Our record of climates in the past can be extended using geological evidence. For example, warm climates will favor the growth of corals at low latitudes while colder climates promote the formation of glaciers at high latitudes. Both leave their signatures in rocks, as will be expanded upon later in this chapter. Magnetism also contributes to our understanding of paleoclimates in two different ways. The first involves the type and properties of the magnetic minerals found in different locations, and the second involves paleomagnetic research using the record of Earth's magnetic field stored in rocks (chap. 2).

The first of these, referred to as environmental magnetism, does not use the fossil magnetic record. Instead it uses the magnetic properties of rocks, such as induced and remanent magnetizations (chap. 1) produced in a scientific laboratory, to obtain information on past environments, including paleoclimates.

Environmental magnetism is founded on rock magnetic measurements, such as carried out by Subir Banerjee, David Dunlop, and Mike Fuller, world-class rock magnetists who perform experiments to determine the magnetic properties and origins of minerals and rocks. Of these, only Banerjee has received the Fleming Medal; he also considered leaving science to become a historian, taking a sabbatical leave at the University of California at Berkeley, where he taught himself Arabic to learn more about the history of science. Rock magnetism involves the determination and explanation of the magnetic properties of minerals and rocks. While rock magnetism is often viewed as a specialized subject within paleomagnetism, it requires a basic understanding of physics, chemistry, and geology. It also is the basis of environmental magnetism. Rock magnetism, introduced in chapter 2, is expanded on in the appendix.

Environmental magnetism is something of an eclectic science that involves the use of any magnetic property to provide information on the environment. The first clear use of environmental magnetism occurred in 1926 when Gustav Ising used magnetic susceptibility to show that laminations in lake sediment in Sweden formed because the ice margin source of the sediment was closer in winter than in summer. Sediments deposited close to the ice margin were found to have a different magnetic susceptibility from those deposited farther away. A different example of environmental magnetism involves industrial pollutants. Because they often contain iron oxides and sulfides, they contribute magnetic materials to sediments downwind from the place where they were emitted.[12] By measuring the magnetic properties of sediments, scientists can obtain some understanding of when the effluents were emitted and the extent of the pollution. Another example involves the subject of biomagnetism, the ability some animals have to make magnetic minerals. For instance, magnetotactic bacteria manufacture magnetite and use it to sense Earth's magnetic field (chap. 5), and chitons, a species of mollusk often found in tide pools, make magnetite teeth that are sufficiently hard to allow them to scrape food off rocks. Environmental magnetism became widely recognized in 1986 after Roy Thompson and Frank Oldfield published an entire book devoted to the subject.[13]

While this is not the place to review environmental magnetism in its entirety, I will illustrate its usefulness in paleoclimate studies. You might have wondered why dust, particularly a type called loess, seems to vary in ice cores with the Milankovitch cycles, as does temperature and carbon dioxide (fig. 6.1). A pioneer of geology, Charles Lyell first described yellowish-gray dust deposits in Europe and North America as loess (from the German *Löss*, meaning "loose"). Glaciers sometimes grind down rocks into a fine powder, called rock flour. After drying, rock flour is easily eroded and transported by wind. It is deposited to form sedimentary rocks called loess. The amount of loess deposited varies with glacial cycles, which in turn are correlated with the Milankovitch cycles.

Not all loess deposits are derived from rock flour. Lyell never saw the most spectacular loess deposits, which form the loess plateau (also called the Huangtu Plateau) in north-central China. This plateau covers some 640,000 acres and has yellow dust deposits that typically exceed 100 meters. This loess is derived from wind-blown dust eroded from desert sediments. A German scientist, Friedrich Heller, paired up with a Chinese scientist, T. S. Liu, during the 1980s to show that magnetic susceptibility could be used to map out the sources of the dust. Rock magnetists such as Banerjee had already shown that magnetic susceptibility depends on many factors in addition to mineralogy, such as grain size, shape, and even defects in the crystalline structure of the magnetic minerals (appendix).

Therefore, magnetic susceptibility can be used as a sensitive parameter to characterize the source of materials, including the loess in China. These loess deposits were laid down by monsoons during the past 2.5 million years, the modern ice age. They were strongest in glacial intervals when the winds arrived from the northwest and weakest during interglacial intervals when the monsoonal winds blew from the southeast. The loess plateau consists of many layers of loess alternating with fossil soil deposits (called paleosols). The loess was deposited during colder arid glacial intervals while the soils were formed during warmer wet interglacial intervals. Heller and Lui found, as was verified and extended by others, that the unaltered loess has a much smaller magnetic susceptibility than does the fossil soil, probably due to chemical changes accompanying the soil formation. Environmental magnetists, including Heller and Liu, used magnetic susceptibility to determine the monsoonal history of the past 2.5 million years of north-central China.

This is not the only example where magnetism has proved useful in the determination of our planet's glacial history. The use of magnetism has also helped to define the most dramatic climate changes this planet has ever seen: at times during the Precambrian (prior to 544 million years ago), the entire planet was completely covered by ice or nearly so. During the Precambrian, the continents looked entirely different, the ocean currents were different, and the composition of our atmosphere was different. To appreciate this, we need first to understand how global tectonics can affect climate.

*   *   *

As science evolves with time, so do the views of experts. Using the language of the philosopher Thomas Kuhn, a paradigm shift occurred during the 1960s in earth sciences with the acceptance of plate tectonics. Instead of simply referring to the movement, or drift, of continents, we refer to the movement of plates, pieces of the upper part of Earth, which often include both continental and oceanic crust that move as individual units. The plate tectonics model has now become the basis for understanding the major geological processes acting on Earth. Therefore, you may be surprised when I tell you it sometimes fails. But I am getting ahead of myself.

The upper "rigid" part of Earth, called the lithosphere, consists of seven major plates (in terms of area), and many minor plates that slowly creep across the face of Earth (fig. 6.2). Although the lithosphere varies in thickness over the surface of Earth, on average it is about 100 km

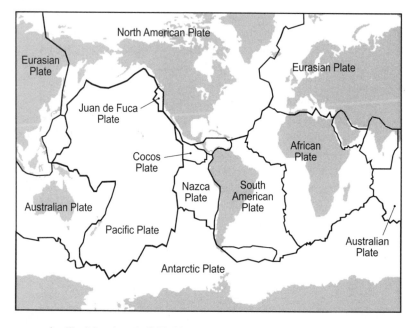

FIGURE 6.2 The lithosphere is divided into many plates. Plate boundaries are indicated by thick black lines. Examples of major plates are the Pacific Plate and the North American Plate. Examples of minor plates are the Juan de Fuca, Caribbean, and Cocos plates. Not all the names of the minor plates are given in this diagram. Figure drawn by Beth Tully.

(62 miles) thick; it consists of the crust and the uppermost part of Earth's mantle (chap. 3). When the plate tectonics model was formulated during the 1960s, plates were described as being internally rigid pieces of the upper part of Earth with three types of boundaries. Divergent boundaries, most commonly described as spreading centers or mid-ocean ridges, are locations on Earth where two plates are pulled apart. There magma rises up and adds new lithosphere to the plates. While lithosphere is "created" at spreading centers, it is "destroyed" at the second type of boundary where plates converge. Convergent plate boundaries are identified by their associated subduction zones, in which earthquake foci descend into Earth at an average angle close to 45° (fig. 6.3). Sometimes the lithosphere becomes so dense from contraction during cooling that it breaks under its own weight and sinks into the mantle to initiate a subduction zone. As it sinks into the mantle, the lithosphere pulls the trailing part of the plate away from the spreading center. The third type of plate boundary is the transform fault: it is a boundary in which new material is neither added to,

nor subtracted from, a plate. Two plates grind horizontally past each other at transform faults. This movement can produce earthquakes, such as occur along the San Andreas Fault in California. The San Andreas Fault is the best-known transform fault in North America: it marks one of the boundaries between the Pacific and North American plates (fig. 6.2).

I am sometimes asked, "Who discovered plate tectonics?" I reply that many scientists were involved and no one or two individuals should be given credit for the discovery. Although a few of these scientists were mentioned in chapter 2, many others were not. One not mentioned is the Canadian J. Tuzo Wilson (1908–1993), who discovered transform faults. Wilson was a fascinating character who told me he liked to "shake people up." He also had a great sense of humor, which was exhibited in a speech delivered after being elected president of the American Geophysical Union (AGU), which,

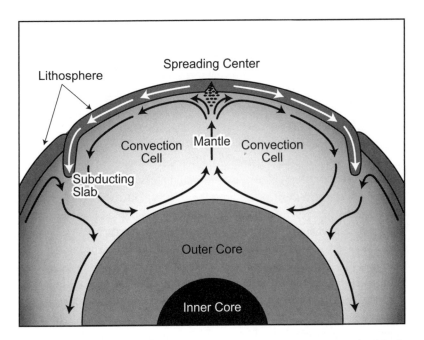

FIGURE 6.3 A simplified picture of the lithosphere's movement and mantle convection. Mantle plumes, which give rise to hotspots, are not shown. The figure emphasizes that plates are formed where mantle upwelling in a convection cell occurs. Downward mantle flow occurs where plates descend. Although no reliable earthquake focus has been recorded at a depth greater than 700 km (435 miles), some subduction zones are now recognized to extend to much greater depths. This is manifested in seismic tomography models, which, like the use of ultrasound to image an unborn fetus, provide images of Earth's interior. Seismic tomography models also indicate that mantle convection is far more complicated than shown. Figure drawn by Beth Tully.

with more than 50,000 members, is the largest earth science organization in the world. He said he was proud to be elected president of the AGU in 1978, particularly considering that every manuscript he had submitted to an AGU journal had been rejected. He typically arrived at his scientific conclusions through intuition rather than mathematical analysis, which apparently allowed reviewers to recommend rejection of his manuscripts. As president of the AGU during the Cold War involving the United States and the Soviet Union, the Soviets incorrectly thought that Wilson had top-secret information on American seismological research related to distinguishing underground nuclear bomb explosions from earthquakes. Over dinner he once told me of an encounter he had while visiting Russia. Upon returning to his hotel room after consuming considerable amounts of alcohol at an official dinner, Wilson opened the door to find a beautiful woman lying naked on his bed. The woman jumped up, ran to him, and started kissing him. During this brief interval of time, numerous flashes went off, indicating cameras were recording the event. Once back home at the University of Toronto, Wilson received a letter containing some pictures taken in the Russian hotel room. The letter said "they" (presumably members of the Russian secret service, then the KGB) expected Wilson to be cooperative on various scientific issues in the future. After showing the pictures around, including to his wife, Wilson wrote on the back of one of the pictures something like "Thanks for showing me a good time in Russia" and sent the picture back to the sender, who apparently never bothered him again.

Wilson made several contributions to plate tectonics in addition to discovering transform faults, including the proposal in 1963 that some volcanism is associated with fixed hotspots that could be used to determine the lithosphere's motion. In 1971 W. Jason Morgan further developed this hypothesis to suggest that hotspots originated close to the base of the mantle. Thus, Morgan reasoned, they would be fixed relative to each other and could be used to provide an absolute reference frame to describe the motion of plates. A classical example of a hotspot lies beneath Hawaii, the largest island in the Hawaiian-Emperor seamount chain. The Hawaiian Islands and seamounts extend northwesterly about 3,000 km from Hawaii to the Kure Atoll. At this atoll, the chain of seamounts bends in a more northerly direction along the Emperor seamount chain, which extends almost another 3,000 km (fig. 6.4).

The lava (basalt) flows on the "Big Island" of Hawaii do not exhibit any evidence of reverse magnetic polarity. The entire island has been built

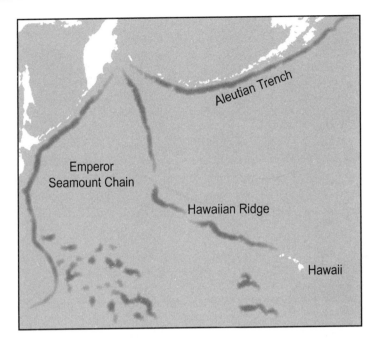

FIGURE 6.4 The Hawaiian-Emperor Seamount Chain extends from the island of Hawaii in the middle of the Pacific Ocean to the western end of the Aleutian Islands. It is a classic example of a hotspot chain. Another example (not shown) is the Society Islands in the South Pacific, which include Tahiti, Mooréa, and Bora-Bora. There are about 50 recognized hotspots around the globe. (These differ from the so-called ring of fire volcanoes around the Pacific Ocean, which are associated with sources originating in subduction zones.) Figure drawn by Beth Tully.

since the last reversal occurred 780,000 years ago (chap. 2). It is still growing, as evidenced by eruptions from Kilauea and Mauna Loa volcanoes, two of the five volcanoes that gave birth to the island. Radiometric dating of samples from other islands and of samples obtained during an international oceanic drilling program indicates that the ages of the first eruptions along the chain increase to the north. The Meiji Seamount, which lies at the northernmost end of the Hawaiian-Emperor seamount chain close to the Aleutian Islands, has an age of 82 million years. The bend in the middle of the seamount chain (fig. 6.4) began 50 million years ago and was completed 42 million years ago. Wilson and Morgan's hypothesis indicates that the Pacific plate moved southward over a fixed hotspot over the past 82 million years. It took a more eastwardly direction between 50 and 42 million years ago.

While hotspots still provide a useful reference frame, not all of them remain fixed, as has become apparent during the twenty-first century. John Tarduno at the University of Rochester and his colleagues tested the fixed hotspot hypothesis by using paleomagnetism. If a hotspot is fixed, all eruptions from it should have occurred at the same latitude, now occupied by Kilauea on Hawaii. He demonstrated that the fixed hotspot model failed by measuring the magnetic directions of many samples from widely separated seamounts along the Emperor chain. The paleolatitude (chap. 2) increased systematically in the northern direction. The hotspot had moved in the opposite direction at a rate around 4 cm per year between 80 and 47 million years ago—a rate comparable to the rate many plates move. His conclusion was later supported by theoretical studies conducted at Harvard University by Richard O'Connell and his associates, who have developed models of mantle convection.

Mantle convection refers to the slow creep of mantle material driven by temperature differences between the deep mantle and the lithosphere. In 1929 Arthur Holmes postulated that mantle convection drove continental drift. The convection of mantle material is 10 million times slower than the convection in the liquid outer core that sustains Earth's magnetic field (chap. 3).

Mantle convection occurs in a "solid" mantle. Seismic shear waves, which cannot propagate in a liquid, travel through the mantle (chap. 3). It may seem odd to refer to convection of a "solid" material. It is similar to solid ice flowing when it is part of a glacier. Some materials act like a solid over a short time interval, but act like a liquid over a longer time interval. For example, a ball of silly putty bounces when dropped onto a hard surface, but it flows if left to rest on that surface for a few hours. Although the ball behaves elastically during the short time it bounces, it flows like a fluid over a much longer time interval. Similarly, the mantle behaves like an elastic solid for the time it takes an elastic wave to pass by (seismic, or elastic, waves traverse the entire mantle in less than an hour), but it behaves like a liquid over the course of a year. The Earth's "solid" mantle creeps at a rate of a few centimeters per year (much slower than most glaciers flow) to drive plate tectonics (fig. 6.3). Hotspots rise up through this creeping mantle as relatively less dense "solid" plumes. They remain "solid" in the conventional sense until they get within about 100 km of Earth's surface, where some melting occurs. Eventually, just as a warm and shaken can of soda erupts when opened, molten material, magma, erupts at Earth's surface to form a hotspot volcano.

Although Wilson's hypothesis of fixed hotspot sources has not been confirmed, it has provided valuable stimulation to science. Moreover, the motion of hotspot sources relative to plates is used today to provide valuable information on the character of mantle convection.

On a clear day, the glacier-clad Olympic Mountains can easily be viewed across Puget Sound from the University of Washington campus. The Olympic Peninsula, on which these mountains sit, was not part of the North American Plate 50 million years ago. For that matter, much of Washington and western British Columbia were not part of the North American Plate 100 million years ago. The plate tectonic model discussed above does not explain their origins. Geologists are increasingly using the phrase "global tectonics" because of modifications made to the strict plate tectonic model developed about four decades ago. One important modification is that subduction zones occasionally jump to different locations. Much of Washington state and western British Columbia consists of displaced terranes. These terranes were sections of lithosphere that formed on other plates and later transferred to the North American Plate. Perhaps as many as fifty instances have occurred when the subduction zone jumped westward from the edge of the North American Plate during the past 100 million years. Each time a terrane was added, the boundary of the North American Plate jumped westward.

A micro-continent, or even a section of plate carrying a large seamount, cannot always be subducted. It can be too thick and buoyant to be carried down along with the oceanic lithosphere. When a micro-continent encountered the western boundary of the North American Plate, it halted the previously active subduction. This increased the stress on the lithosphere, eventually causing the plate to break on the west side of the micro-continent. A new subduction zone was thus created, making the previous subduction zone on the east side inactive. When the most recent terrane, the Olympic terrane, was added to Washington state about 15 million years ago, it pushed up the Olympic Mountains. The tall mountains to the west of Seattle were not there prior to about 15 million years ago.

Displaced terranes are now recognized to occur in several places around our planet. Alaska is essentially made up of various slivers of material delivered to the North American Plate from other plates that traveled northward to transfer parts of their lithosphere to Alaska. Evidence for this comes from paleomagnetic measurements yielding paleolatitude information (chap. 2). Older rocks have more shallow magnetic inclinations, indicating that they were formed at lower latitudes than younger rocks. Similarly, India traveled a vast distance from the Southern Hemisphere to collide with the rest of Eurasia about 50 million years ago. In this case, plate collision did not initiate subduction; instead it gave rise to the Himalayas and the Tibetan Plateau. Other processes produce deformation within the plates themselves.

In 1811 and 1812, some of the largest earthquakes ever known to occur away from plate boundaries struck near New Madrid, Missouri (then the Louisiana Territory). They were felt over nearly 50,000 square miles. They substantially altered the terrain, including forming new lakes and changing the course of the Mississippi River. Missouri, which lies near the middle of the North American Plate, is not where large earthquakes are supposed to occur: they are supposed to be confined to plate boundaries.

The above are only a few of many examples illustrating the failure of the original plate tectonics model. Using magnetic data, modern analyses indicate that all three types of plate boundaries occasionally jump. In addition, on rare occasions transform faults have allowed magma to erupt along their boundaries. (These are called leaky transform faults.) The lithosphere near Indonesia appears to be a region in which rearrangements of plate boundaries are presently occurring.

You may be puzzled as to why we still teach plate tectonics in beginning geology courses (including ours at the University of Washington), when such exceptions to the plate tectonic model are well documented. Scientists do this because plate tectonics is an excellent (although imperfect) approximation of global tectonics. For example, although the occurrence of the great New Madrid earthquakes is well established, the vast majority of large earthquakes occur on plate boundaries. More than 90 percent of all the energy released in earthquakes occurs in subduction zones. The rest of the energy released primarily occurs from earthquakes located along transform faults and spreading centers. Large earthquakes that occur away from plate boundaries, such as the New Madrid ones, are relatively uncommon. Although deviations from the strict plate tectonics model cause changes in the evolution of plates and their boundaries, most changes require many tens of millions of years to be easily recognizable. However, exceptions occur and this creates difficulties in reconstructing paleoclimates.

Many factors affect climates on a geological time scale, including the composition of our atmosphere, the locations of continents, and the shifting of oceanic currents. The use of magnetic stripes (magnetic anomalies; see chap. 2) and paleolatitudes obtained from paleomagnetic measurements (also discussed in chap. 2) are the primary quantitative tools allowing scientists to reconstruct past locations of plates and continents. Using such information, scientists also reconstruct possible locations of past oceanic currents. For example, 56 million years ago, at the beginning of the geological period called the Eocene, Australia was connected to Antarctica. Warm equatorial waters mixed with the colder Antarctic waters, producing uniform and high temperatures around the globe.

But this is by itself insufficient to explain why a peak in temperature, more than 6°C above our present average global temperature, occurred at the onset of the Eocene.[14] These are the hottest temperatures Earth has experienced during the Cenozoic era, the interval of time since the demise of the dinosaurs 65 million years ago. The duration of these high temperatures is somewhat uncertain, but is estimated to be around 100,000 years. It caused the extinctions of many species. The reasons for the relatively sharp temperature rise are disputed. Perhaps gas hydrates, called clathrates by specialists, were involved. In particular, methane, a much more potent greenhouse gas than carbon dioxide, is sometimes caged within frozen water at depth in sediments. This frozen water–methane mix is called a methane clathrate. Perhaps as warming occurred 56 million years ago, large clathrates melted to release methane, which produced a sharp increase in global temperatures through the greenhouse effect.

Unfortunately, scientists' ability to determine plate locations generally decreases further back in time. One reason for this is that the geometry of plates and their locations becomes more difficult to estimate reliably. Magnetic stripes recorded in the oceanic crust can only be used to estimate plate evolution back to 160 million years, because almost all older oceanic crust has been subducted. (There are a few rocks in the western Pacific that have ages around 170 million years BP.) Rocks are heated to sufficiently high temperatures during subduction that they lose their primary magnetization (chap. 2 and appendix). Hence, all fossil magnetic evidence stored in the oceanic crust is lost during subduction. Quantitative information on the location of continents prior to 160 million years relies mostly on the magnetic inclinations obtained from primary magnetizations of continental rocks. This information yields estimates for paleolatitudes, but not for paleolongitudes. (However, paleomagnetic records of declination can be used to infer the orientation of the site; for example, whether the location rotated with respect to the geographic pole.) In addition, over time many continental rocks are deformed by stress, reheated, and chemically altered. This makes it difficult, and sometimes impossible, to retrieve their primary magnetic record.

To minimize problems associated with the alteration and destruction of primary magnetization, paleomagnetists rely on rocks obtained from the cratons of continents. A craton is an old and stable part of a continent. Sections of the craton's crust have undergone little alteration during the drift of continents. Unfortunately, errors still occur and accumulate in older rocks. Statistics becomes an important tool in such cases. Paleomagnetists are often heard to say something like, "The most probable plate reconstruction is the following . . ." Although most plate reconstructions in the Precambrian (before 544 million years ago) are speculative, there is broad agreement that a supercontinent, called Rodinia, existed a billion years ago.[15] It broke apart about 750 million years ago. During the 150 million years following the breakup, Earth occasionally experienced some of the coldest climates it has ever known. Some scientists even have suggested Earth's surface has occasionally been completely frozen.

*   *   *

Thirty-four million years ago, the climate began to cool and glacial ice started to build up in Antarctica. However, the modern ice age did not begin until 3 to 2.5 million years ago. Since then, there have been many glacial and interglacial intervals, including the Holocene, the present interglacial interval initiated 12,000 years ago. Because interglacial intervals only last about the present extent of the Holocene, the 20 years of cooling in the mid-1970s led some scientists then to speculate that Earth was only a few hundred years away from entering another glacial interval. Reporters and talk show hosts occasionally still repeat this speculation as evidence that atmospheric scientists' forecasts of global warming are unreliable. Naturally when this is done, it is not mentioned that such speculations lacked the endorsements of such organizations as the American National Academy of Sciences or the American Geophysical Union.

The causes of ice ages are a matter of conjecture. The first known one occurred about 2.5 billion years ago. Others occasionally followed. The ice age preceding the present one occurred about 300 million years ago when the supercontinent Pangaea existed. Pangaea lasted until 175 million years ago, when it divided into two smaller supercontinents, Gondwana (the southern one) and Laurasia (the northern one). By then the ancient glaciers had long disappeared and dinosaurs were present.[16] The 300-million-year-old glaciers were found only in places that had been part of Gondwana: Antarctica, Australia, Africa, and India (which at that time was far to the south of the rest of Eurasia). Glaciers were confined to Southern Hemisphere landmasses, which, with the exception of Antarctica, were closer to the South Pole than they are today.

Although the Milankovitch cycles help to explain variations within an ice age (as discussed earlier in this chapter), they do not explain the initiation of an ice age. The positions of continents, changes in oceanic currents, the effects of volcanoes, and the variations in greenhouse gases are some of the factors that can promote ice ages. For example, one leading contender for the modern ice age involves Central America—or should I say the lack of Central America. In 1965 Sir Edward Bullard carried out the first computer fit of the continents by closing the Atlantic Ocean. Instead of using the continental margins defined by sea level as done by Wegener (chap. 2), he used a 500-fathom (3,000-foot) contour depth on the continental shelf. Many scientists were puzzled as to why Central America was missing in his reconstruction. When I once asked this question of Bullard, he simply replied: "It didn't fit."

Later scientists learned that much of Central America didn't exist until various displaced terranes were recently emplaced. The Isthmus of Panama came into existence about 3 million years ago. Before this, the Central American Seaway allowed the waters of the Pacific and Atlantic Oceans to mix freely. The Pacific Ocean has a lower salt content (it has a lower salinity) than the Atlantic Ocean, and this affects ocean circulation: greater salinity increases the density of water, which makes it try to sink below less dense water. The closure of this seaway must have had a major impact on climate. However, it is still debated whether the formation of the land bridge between North and South America is by itself sufficient to have caused the world to plunge into a new ice age.

Based on evidence of glaciations in Precambrian rocks on all continents, Brian Harland (1917–2003), a professor at Cambridge University in England, proposed in 1964 the existence of a Precambrian ice age. Although initially this proposal was not well received, within a few years he had convinced the geologic community of the reality of Precambrian ice ages. About a decade later, paleomagnetists, such as my colleague Mike McElhinny, puzzled over evidence suggesting that some Precambrian glaciations with ages between 750 and 560 million years ago seemed to have formed at low latitudes.

McElhinny, who was born in India and completed his PhD degree in space physics at the University of Rhodesia, spent the majority of his career in Australia. He was a genius when it came to data, and for many decades he managed the paleomagnetic database for the international paleomagnetic community. I recall one time when he called me into his office at the Australia National University (ANU), saying, "Look at this table published by Bob" (not his real name). The table contained about twenty different data entries and looked fine to me, and I said so. Pointing to a single entry in the table, he said, "Don't you see this is not the same number he used three years ago?" "You've got to be kidding, Mike," I replied. "Even you couldn't remember that." He grumbled something in reply, went over to his vast paper collection, retrieved Bob's article from three years earlier, and showed me he was indeed correct.

I often heard McElhinny puzzle over possible reasons paleomagnetic data taken from rocks associated with Precambrian glaciations had shallow magnetic inclinations. This suggested that glaciers might have occurred at very low latitudes. Although geologists recognized mountain glaciers were possible in the tropics, such as exist today on Mount Kenya on the equator, the formation of large-scale continental glaciers were thought to be restricted to high latitudes. The evidence that these ancient glaciers were extensive can be found in certain rocks and landforms. For example, most of the rocks used to indicate the presence of Precambrian glaciations are called tillites, rocks that form from the compaction of till. Till consists of a jumble of rock sizes, ranging from small grains to boulders, formed by the grinding action of glaciers.

Although tillites were used to argue for the presence of glaciers in the Precambrian, they were not used to obtain paleomagnetic data. The erosion and transport of rocks by glaciers distorts till too much to obtain a record of the paleomagnetic field from it. Instead, paleomagnetists used other rocks that appeared to have formed about the same time as the tillites. This offered a possible way to solve the dilemma of Precambrian glaciers forming at low latitudes. Some paleomagnetists posited that a continent, for example, Africa, was glaciated when it was at high latitude and subsequently moved rapidly to low latitude, where it recorded a shallow magnetic inclination when new non-glaciated rocks formed. I recall McElhinny once complaining to me that some paleomagnetists had Africa bouncing up and down between high and low latitudes like a yo-yo. While McElhinny never solved this puzzle, other paleomagnetists who passed through his laboratory at ANU did.

McElhinny's vast knowledge of paleomagnetism and tectonics drew many paleomagnetists to ANU. One of these was Brian Embleton, who spent several years with McElhinny during the 1970s. Although born in Great Britain, Embleton fits the image of an Australian geologist: a strong, rugged no-nonsense-type of individual, who later showed he had administrative talent by successfully managing many Australian government science programs. After leaving ANU, he joined the CSIRO (Commonwealth Scientific and Industrial Research Organisation) in Sydney as a geologist and eventually hooked up with George Williams, who was studying Precambrian glaciations in the Flinders Ranges in southern Australia. They published a paper concluding the tidal rhythmites studied by Williams (and discussed earlier in this chapter), which had been deposited during a Precambrian glaciation, contained a nearly horizontal primary magnetization. This implied that these rocks had formed when southern Australia was within a few degrees of the equator.

Joe Kirschvink, a geologist and biologist whom we met in the previous chapter, was convinced that the Embleton-Williams result must be wrong. I first met Kirschvink about three decades ago while he had a postdoctoral position under McElhinny, and I have kept in contact with him ever since. He is one of the most creative and provocative scientists I know. What he lacks in physical stature, he makes up for in energy. He seems to thrive on controversy and always brings new arguments to any scientific discussion. As an aside, his wife lives in Tokyo, where their children attend school. Kirschvink regularly commutes between Tokyo to see his family and Los Angeles, where he is a professor at Caltech.

Kirschvink was aware of a model developed during the 1960s by a Russian atmospheric scientist, Mikhail Budyko, who had concluded that if an ice age involved large glaciations extending below 30°, the entire Earth would freeze. Budyko's model is based on climatic changes associated with albedo, the fraction of incident light reflected from a surface. Albedo varies for different conditions. For example, it varies depending on the character of water waves or whether there is new snow on a glacier. However, on average the albedo of glaciers is about eight times that of water. When light is not reflected, it goes into heating Earth. Budyko found that if there was much ice cover at latitudes below 30°, Earth cooled and more ice formed. As more ice formed over previously ice-free regions, more light was reflected and Earth cooled further. The more Earth cooled, the greater the ice cover became, a positive feedback. When this occurs below 30° latitude, it produces a runaway effect leading to a completely ice-covered planet. Budyko had concluded this never happened because once frozen Earth would forever remain in a frozen state. Kirschvink was convinced this showed that Embleton and Williams were wrong. But where was their error?

After thinking about this for a while, Kirschvink concluded that the magnetization in the southern Australian rocks did not accurately record Earth's ancient magnetic field direction. He was aware that some of the rocks examined by Embleton and Williams had been squeezed and folded after the rocks had formed. He had an undergraduate student, Dawn Summers, make magnetic measurements on some of the folded rocks to demonstrate that his friend and fellow paleomagnetist Embleton had not measured a primary magnetization, a magnetization acquired when the rock formed. Kirschvink was convinced that the magnetization in these samples would be a secondary one acquired when Australia was at high latitudes—long after the rocks had formed. Later, or so Kirschvink conjectured, Australia drifted to near equatorial latitudes, where some process, such as chemical alteration, allowed the magnetic record to be reset (chap. 2 and appendix). He recommended that Summers test this conjecture by using the paleomagnetic fold test developed by my colleagues McFadden and McElhinny. If the primary magnetization had not been altered when tectonic stresses folded the tidal rhythmites, it would remain parallel to the sediment layers, which originally had been deposited horizontally in a shallow water environment. However, after folding, some of the layers were no longer horizontal. If the magnetization had been acquired after the rocks were folded, it would not be parallel to the layering.

Summers carried out the test and found the magnetization was a primary one—contrary to Kirschvink's expectation. Subsequently other paleomagnetists—such as Phil Schmidt, who also worked several years with McElhinny—have conducted additional consistency tests and confirmed that the magnetization measured by Embleton was indeed primary. Kirschvink concluded that there was little doubt that extensive glaciers occurred in the Precambrian when southern Australia was close to the equator.

After puzzling over this unexpected finding, Kirschvink hit upon a new idea. Because Budyko failed to take into account volcanoes, his conclusion that Earth could not recover from a completely frozen state was incorrect. Volcanoes such as Mount Rainer and Mount Saint Helens in Washington are called stratovolcanoes or composite volcanoes. They are typically steeper than the volcanoes that make up Hawaii (called shield volcanoes because they are shaped like a shield). Composite volcanoes are so named because they consist of both lava flows and explosive material. Although their lavas exhibit a much wider range of chemical composition than hot-spot volcanoes, their average composition is that of andesite, which has more silicon and less magnesium and iron than does the basalt of a hot-spot volcano. Composite volcanoes, which occur above subduction zones, make up the "ring of fire" around the Pacific. They are formed when water and gases, such as carbon dioxide, are released at a depth between 100 and 150 km from a descending subduction zone. Because water lowers the melting temperature of most materials, some melting occurs after water is released into the mantle from the subduction zone. This eventually leads to the eruption of magma in the form of volcanoes, which lie 60 to 90 miles above a subduction zone. Carbon dioxide is released into the atmosphere during these eruptions. Kirschvink thought the buildup of carbon dioxide over a long time in our atmosphere from volcanic eruptions could produce a super greenhouse effect and melt the ice covering Earth.

Water is required in the atmosphere to transport carbon dioxide to Earth's surface. During the ice age, it was too cold for much water to evaporate from the ice. There would have been little rain available to transfer carbon dioxide to Earth's surface, where it could have combined with magnesium and calcium to form carbonate rocks, a process that has been ongoing throughout much of geological time. Thus carbon dioxide continued to build up and heat the atmosphere during the ice age. Eventually Earth's surface became hot enough to initiate melting of the ice. This produced increased vaporization and higher water content in the air. Because water is an important greenhouse gas, this led to further heating of the atmosphere, a positive feedback. The rate of glacial melting further increased, bringing an end to the ice age.

Kirschvink thought that at the end of this ice age, the water in the atmosphere would have been sufficient to transfer substantial amounts of carbon dioxide in rain to Earth's surface, where it could be used to form large volumes of carbonate rocks (dolomites and limestones). This possibility is supported by the observed abundance of carbonate rocks that formed at the end of the ice age. Indeed, the volume of the carbonate rocks formed at that time is so high that Kirschvink concluded that there must have been a considerable amount of carbon dioxide in the atmosphere at the end of the ice age. Years later theorists, such as Dick Peltier at the University of Toronto, would conclude that about 350 times more carbon dioxide than is presently in our atmosphere was required for Earth to exit a completely frozen state—one in which glaciers covered the land and sea ice blanketed the oceans. Kirschvink labeled such a frozen state "Snowball Earth."

Although Kirschvink likes to promote big ideas, in this case he seems to have been a bit unsure of himself. He did not publish a paper on it until four years later in 1992 in a chapter of an edited book.[17] It appears to not have been widely read or appreciated until much later. Even though I often had dinner and coffee with Kirschvink, I was unaware of the idea of Snowball Earth until two Harvard scientists, Paul Hoffman and Dan Schrag, extended and publicized it in papers and talks.

Hoffman was doing geological work in Namibia and was puzzled by cap carbonates, a thick layer of carbonate rock. He noticed cap carbonates immediately overlying glacial deposits in Namibia. This was not atypical. Cap carbonates seem to commonly lie above the roughly 635-million-year-old glacial deposits elsewhere in the world. They are often meters thick and in some places tens of meters thick. Hoffman wondered what caused a thick section of carbonate rock to form immediately after the Precambrian ice age ended. He often discussed this problem with his colleague Schrag. During one of these discussions, he brought forth some ideas he had learned from sitting next to Kirschvink at a dinner in 1989 at a scientific meeting.[18] The more they discussed Kirschvink's hypothesis, the more likely they thought Snowball Earth actually occurred. But how could they test it?

Unlike carbon-14, the radioactive isotope used for dating, carbon-13 and carbon-12 do not decay over geological time. Carbon-13 differs from carbon-12 by having an extra neutron. This makes it heavier and less desirable for cyanobacteria to use in photosynthesis,[19] thus cyanobacteria have a slight preference for carbon-12 over carbon-13. Therefore, the ratio of carbon-13 to carbon-12 in cyanobacteria is lower than in the atmosphere. When these ocean-dwelling photosynthesizers die, their organic remains contribute carbon with a larger ratio of carbon-12 to carbon-13 to the sediments accumulating on the ocean floor than was in the oceanic water. Hoffman and Schrag were aware that a substantial amount of carbon isotope data from rocks were available after the termination of the Precambrian ice ages. They thought the isotope data should reflect a decrease in photosynthesis at the onset of a major ice age and a rebound after the ice age ended. They found this prediction to be correct: the isotope data are consistent with the Snowball Earth hypothesis. For example, a sharp downturn in the ratio of carbon-12 to carbon-13 is recorded in sediments at the onset of the 635-million-year old ice age that Hoffman was studying. They concluded that the Snowball Earth hypothesis appeared to explain all the available geological data related to the subject, including some I have not discussed,[20] and published a widely read paper on Snowball Earth in 1998. Almost immediately, Hoffman began to give numerous talks on the subject to publicize the idea. They also followed with additional papers, as did many others.

As scientists are skeptical of any idea they do not think of first, objections to the Snowball Earth hypothesis quickly emerged. Most scientists now accept that extensive glaciations reached down to the tropics four or more times during the Precambrian. But many don't find the evidence convincing that Earth's continents and oceans were completely frozen. They question whether primitive life could have survived such harsh conditions. They argue that some stromatolites, fossil remains of cyanobacteria, indicate that cyanobacteria did not seem to suffer much during the major Precambrian ice ages. Many scientists now favor a modified version of the Snowball Earth hypothesis, dubbed the "Slushball Earth" hypothesis. In this modified version, Earth's entire surface was not frozen over. Parts of the ocean were either ice-free or the ice cover was very thin. This allowed photosynthesizing organisms to survive. Other scientists have even suggested that some regions on land were ice-free due to the lack of precipitation. In any case, Kirschvink's Snowball Earth hypothesis or its modified version, the Slushball Earth hypothesis, is one of the handful of most stimulating hypotheses advanced in earth science during the past two decades.

Epilogue: Some Parting Comments

After I finished writing a draft of this book, a newspaper article appeared saying scientists have used Google Earth, a useful Internet program that allows one to view essentially any location on Earth's surface, to show that cows (and some species of deer) align along magnetic field lines. The authors conclude that cows sense the magnetic field. They have subsequently supported their arguments by providing evidence that cows do this except when they are near power lines, which are supposed to disrupt their magnetic sense. Should I revise the book to include this? Although published by reputable scientists, their claim reminded me of one made by the journalist Paul Brodeur, who wrote a series of articles in the *New Yorker* in 1989 suggesting that people who lived near power lines were more likely to get cancer. This claim led to many lawsuits, even after physicists presented evidence suggesting this was pseudoscience.[1] I read the papers on cow (and deer) orientation to learn more.[2] Although the statistics given looked impressive, I had questions about some of the details of their calculations. I also wondered why cows would choose to orientate with respect to the magnetic field rather than in a direction to affect the heat they receive from the Sun or to minimize the effects of wind. Can cows be conditioned to sense a magnetic field? Upon reflection, I decided not to add a section on cow magnetism to chapter 5. Other less controversial topics have also not been included in this book. For example, I chose not to discuss the induction of electric currents in our crust and mantle from magnetic field variations in our ionosphere. These currents provide valuable information on the properties of Earth's interior, including the location of underground water and buried archaeological ruins. In spite of these omissions, I have tried to present a variety of different topics to illustrate the broad impact electromagnetism has had on Earth and its environment.

Perhaps it is not so surprising that magnetism and magnetic fields affect so much of our understanding of our world: the electromagnetic force is one of the four known fundamental forces that alter our physical universe. At the beginning of the nineteenth century, there were three known fundamental forces: the gravitational, electric, and magnetic forces. This was reduced to two at the beginning of the twentieth century because James Clerk Maxwell had shown that the electric and magnetic forces could be combined into one force, the electromagnetic force. Two more forces were introduced in the twentieth century, the strong and weak nuclear forces.[3]

The so-called standard model of particle physics, mostly developed during the 1970s, unifies three of the fundamental forces, excluding only the gravitational force. Today particle theorists are trying to develop a theory to unify all the known forces, once a goal of Albert Einstein's. Such a theory is called the theory of everything, or TOE. Although unification of the four fundamental forces is an admirable and probably even an achievable goal, I doubt there ever can ever be a true TOE. One reason I believe this can be traced to the work of a brilliant mathematician, Kurt Gödel (1906–1938), who produced a famous result referred to as the incompleteness theorem. Gödel, a Viennese wunderkind of logic who suffered from paranoia, became close friends with Einstein shortly after he arrived at Princeton University. Gödel established his reputation in 1931 by demonstrating a fallacy in a conjecture by Bertrand Russell and Alfred Whitehead, famous philosophers and mathematicians of the early twentieth century. Essentially, the Whitehead-Russell conjecture is that all mathematics is derivable from logic. Gödel used pure logic to show any reasonably general mathematical system contains a theorem that cannot be proved or disproved.[4] Consider a logic system in which every statement is required to be true or false. Now consider the statement "This sentence is false." Is the sentence true or false? A contradiction is arrived at regardless of your choice. This illustrates that the logic system is incomplete.[5] I suspect a similar theorem applies to science. That is, science will never be complete: there will always be something more for scientists to discover. Indeed, I believe there are no absolute scientific laws accessible to humans. Even if I am wrong, how would you know? I believe that the best we can do is to converge on laws that better describe our universe. Naturally there are missteps along the way, as I hope I have illustrated by incorporating some history of science into the story.

An appreciation of the gains the science of electromagnetism has made can be obtained by reflecting back to the beginning of the twentieth century, a time when we only knew of two fundamental forces. Even so, we knew a surprising amount. For example, we had Maxwell's famous four equations of electricity and magnetism, which are still used as a starting point for most advanced electricity and magnetism courses in the twenty-first century. Nevertheless, we did not yet know of solar and planetary dynamos, magnetic field reversals, plate tectonics, magnetospheres, Van Allen radiation belts, planetary magnetism, Precambrian ice ages, or that lowly bacteria could sense Earth's magnetic field. There was no concept that electric currents could organize themselves in the interiors of stars and planets to produce global magnetic fields. We did not know these fields sometimes reversed polarity; that reliable records of Earth's magnetic field could be stored in rocks; and that this rock record would play a crucial role in the development of plate tectonics, the major paradigm shift in twentieth-century earth sciences. We did not know that magnetospheres exist and, on Earth, protect us from solar radiation carried from the Sun in the solar wind. For that matter, we did not even know there was a solar wind. Scientists at the beginning of the twentieth century would have been amazed to learn the extent of present-day planetary exploration; some even had doubts it was feasible for a rocket ever to escape Earth's gravitational pull. Who would have expected that paleomagnetism would be crucial to showing the greatest ice ages of all time occurred hundreds of millions of years ago? Finally, who would have thought that animals, such as turtles, have long been using the magnetic field for navigation at sea, when humans have only learned to do this a little over a millennium ago?

I expect similar advances in our knowledge will occur during this century. Already some scientists have suggested using the solar wind to push spacecraft across our solar system, novel ways to use electromagnetism to image our bodies and to promote healing, and the use of magnetic fields as criteria to sort out which exoplanets (planets beyond our solar system) might harbor life. However, I suspect that the biggest findings by scientists during the twenty-first century will be ones not presently anticipated. The study of magnetism, one of the world's oldest scientific endeavors, still produces remarkable results that delight and surprise us. I am glad there is no clear end to those studies.

Appendix: Rock Magnetism
Fundamentals

The material in this appendix is presented at a substantially higher level than in the main body of this book. It presents information on rock magnetism that readers with a more sophisticated understanding of science might desire. Anyone who wants even more than the overview presented here should consider looking at the reference given in note 1, below.

Magnetic Domains

I mentioned in chapter 2 that compass needles used by early mariners sometimes required remagnetization. One way of doing this is to heat a compass above its Curie temperature, where it loses all of its magnetization. Then the compass needle can be cooled down in Earth's magnetic field while the north end is in the Northern Hemisphere. The needle does not have to be pointing precisely to the north. Indeed, it can be pointing in a direction far off that of magnetic north, because its magnetization will be acquired along the axis of the needle.

Shape anisotropy causes the remanent magnetization in the compass needle to always lie in the direction of the needle. In this case "anisotropy" refers to the observation that it is easier to magnetize the needle along the axis of the needle, called an easy direction, than perpendicular to the needle, called a hard direction. In contrast, "isotropy" means all directions are equal. Anisotropy can be illustrated with the help of the top illustration in figure A.1. A rectangular-shaped single domain grain is uniformly magnetized along its longest dimension to the right, as indicated by the large arrow. In reality, the actual magnetization is the sum of many atomic magnetic moments (tiny dipoles) pointing to the right. The head of each atomic moment within the grain is next to the tail of the moment immediately to its right. Therefore, in the interior of the grain, the plus pole of one moment is canceled by the negative pole of the next one. This occurs throughout the grain except at its ends. On the right end of the grain there are plus, or north, magnetic poles, and on the left end there are south, or minus, poles, as shown. These are referred to as bound magnetic poles, as they are attached to the atomic moments immediately inside the surface. These bound magnetic poles produce a magnetic field outside the sample that goes from the north poles to the south poles, as shown by the field lines in figure A.1.

A magnetic field in the interior of the grain connects the north to the south magnetic poles. This field, not shown in figure A.1, is called the demagnetization field because it is opposite to the direction of the magnetization. If the magnetization in this single-domain grain were up (toward the top of the paper), the number of bound poles would increase because the surface area is larger. This means that the demagnetization field would also be larger. To minimize the size of the demagnetization field, grains prefer to be magnetized along their longest dimensions (shape anisotropy)—all other things being equal. The easy direction of magnetization is along the long axis of a grain (to the right in figure A.1), while the hard directions are perpendicular to the easy direction. There are other sources of anisotropy associated with the arrangement of atoms and bonding within the solid, called magnetocrystalline anisotropy, and associated with stress, called magnetostriction anisotropy.[1]

There could be no remanent magnetization without anisotropy, because thermal fluctuations would quickly reduce the magnetization to zero (in the absence of a magnetic field). Consider a rock sample containing many single-domain grains like the one shown in figure A.1. When a very strong magnetic field is applied, say, to the right, every single-domain grain would be remagnetized to the right. After the magnetic field is removed, the rock retains its magnetization: it is said to have a saturation remanent magnetization. Although there will be thermal fluctuations (chap. 2) trying to move the magnetization around, the shape anisotropy usually keeps the

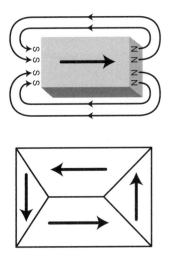

FIGURE A.1   The upper figure shows a rectangular-shaped single-domain grain. The large central arrow represents the sum of the atomic moments within the grain. *N* represents a north
bound magnetic pole, and *S* represents a south bound one; in reality there would be many more
bound magnetic poles than shown. The smaller arrows outside the grain, which go from the
north (positive) poles to the south ones, show the direction of the magnetic field. There is also
an internal "demagnetization" field (not shown) going from the north poles to the south poles
within the grain. This demagnetization field is opposite to the direction of the magnetization.
The lower figure shows a multi-domain grain, one with four domains. The arrows represent
the sum of magnetic moments within a domain. The lines within the grain are domain walls,
which are transition regions in which the magnetic moments (atomic magnets) rotate from
the direction of magnetization in a domain to the direction of magnetization in an adjacent
domain. Figure drawn by Beth Tully.

average magnetization of every grain pointing to the right (exceptions will
be discussed later). In this case, the magnetization is in a minimum energy
state. It would require energy to change the direction of the magnetization
in this grain.

For reasons to be given momentarily, most grains are too large to be
single domain. The four-domain grain shown in the bottom of figure A.1
is an example of a multi-domain grain. The atomic magnetic moments
(magnets) point in the same direction within a domain. The sum of these
moments gives rise to a magnetization within a domain directed along
an easy (magnetocrystalline) magnetic direction. Domain walls are transition regions that separate uniformly magnetized domains. The atomic
magnetic moments at the sides of a domain wall are nearly parallel to the
magnetization in the adjacent domain. These moments gradually rotate
through the interior of the wall to the direction of the magnetic moments

in the adjacent domains, which are not in the same direction. This means every atomic magnetic moment within a domain wall has a component of magnetization directed along a hard magnetic direction.

When a single-domain grain increases in size, the number of bound surface poles increases and so does the energy in the internal demagnetization field. For a large grain, the demagnetization field is too large for the grain to remain uniformly magnetized. In contrast to a single-domain grain, the demagnetization field in a multi-domain grain vanishes. In the four-domain grain shown, there are no bound poles attached to the ends of a domain, because the front end of an atomic moment is directed toward the tail end of the next atomic magnet in line. That is, there is no internal energy associated with a demagnetization field. However, some energy is required to produce and maintain the domain walls.

Smaller single-domain grains do not divide into multi-domain grains because the energy required to produce domain walls is too large. Calculations and observations indicate that the larger the grain is, the larger the number of domains (on average) it contains. For example, when the sides of a cube of magnetite at room temperature exceed about one-tenth of a micron (a millionth of a meter) the grain is no longer uniformly magnetized. A cube of magnetite with dimensions equal to the width of a human hair would be divided into magnetic domains.

A multi-domain grain acquires its magnetization through domain wall movement. For example, the horizontal domain wall in figure A.1 moves upward when a magnetic field is applied to the right. This produces a net magnetization to the right. Domain wall movement typically requires less energy than that associated with the uniform rotation of magnetization within a single-domain grain. Therefore, a single-domain grain typically carries a more stable magnetization (called a hard magnetization in chapter 2) than does a multi-domain grain.

Thermoremanent Magnetization

A thermoremanent magnetization (TRM) is a remanent magnetization acquired by a rock when it cools to room temperature from temperatures exceeding the Curie temperatures of all the magnetic minerals within the rock. It is the primary magnetization acquired by igneous rocks, such as basalts (chap. 2).

Consider an idealized rock in which the magnetic grains are all single-domain magnetite grains with a shape as shown in figure A.1. Further suppose that these grains are randomly oriented. Although each grain exhibits magnetic anisotropy, the rock itself is isotropic. That is, it has no overall easy or hard magnetic directions, even though individual grains do. Rock samples from lava flows, such as basalts, are commonly isotropic, while those from intrusive rocks, such as granite, sometimes exhibit mineral alignment leading to anisotropy. Paleomagnetists test whether a rock is magnetically isotropic by giving it a new magnetization in a scientific laboratory. Isotropic rocks are used to determine the ancient magnetic field directions.

As discussed in chapter 2, the "permanent" magnetization in magnetite decreases from room temperature to the Curie temperature (at 580°C) where it vanishes. This means that the bound poles shown at the ends of the single-domain grain in figure A.1 decrease in size as the temperature increases.[2] Therefore, the magnitude of the magnetic anisotropy increases as the temperature decreases. In other words, the magnitude of the anisotropy barrier separating the two easy directions along the needle (parallel and anti-parallel to the long axis) increases with decreasing temperature. At high temperatures, thermal fluctuations are sufficiently large that the magnetization direction flips back and forth between the two easy directions along the long axis of the grain.[3]

The physics Nobel Prize recipient Louis Néel developed a theory in the late 1940s for TRM. Néel's theory applied to an ensemble (a large collection) of identical particles: all grains were single-domain particles with identical properties, such as chemistry, size, and shape. He found the magnetization was free to move between easy magnetic directions until the magnetic barriers (anisotropy) increased sufficiently with decrease in temperature that the magnetization was "locked in" along one of the easy directions. On average, this magnetization would be in the direction of the applied magnetic field, a minimum energy state. (This occurs because there are more grains on average with magnetization parallel to the external field than in the opposite direction.) Upon further cooling, the locked-in magnetization increases in magnitude, while the average direction of magnetization in a grain usually remains constant.

This last point requires elaboration, because the magnetization is actually not completely locked into place. It can pass over the energy barrier because of thermal fluctuations.[4] Néel calculated a characteristic time, called a relaxation time, that essentially represents the average time it takes for a grain's magnetization to pass over an energy barrier from one easy direction to another. He calculated that this relaxation time grew exponentially as the temperature of the sample decreased. For example, he showed that it was possible for a grain with a relaxation time of one second at 550°C to have a relaxation time of billions of years at room temperature. The magnetization was "locked in" around 550°C and ac- curately recorded Earth's magnetic field direction in an isotropic rock. In this case, 550°C is referred to as the blocking temperature, a temperature that varies depending on factors such as chemistry, grain size, and shape. Above the blocking temperature, the grain is said to exhibit superpara- magnetic behavior. Similar to paramagnetism (chap. 1), the magnetization in an isotropic rock containing only superparamagnetic grains is in the direction of the inducing magnetic field and vanishes when the inducing field is removed.

Néel's TRM theory for single domains has only been altered in minor ways since the middle of the previous century. It remains at the founda- tion of theoretical rock magnetism. Néel also developed a multi-domain theory for TRM around 1955, but this theory has not fared as well as his single-domain theory. Improvements to the theory are ongoing. We will return to TRM theory later in this appendix when we discuss paleointen- sity estimates.

Interpreting the Rock Record

The cartoon in figure A.2 shows a sequence of sedimentary rock layers in the Southern Hemisphere that have recorded Earth's magnetic field. The youngest layer, at the top of the cross section, recorded the present direction of Earth's magnetic field, which points upward and is labeled with the letter H. The second layer, the second youngest sediment in this cross section, points downward. This indicates a reversal of Earth's mag- netic field occurred in an interval of time separating the deposition of the second and the uppermost layer. That is, these sedimentary layers did not continuously record Earth's magnetic field. The situation in the cartoon is exaggerated, as it shows many gaps during the sedimentation process, during which other magnetic field reversals occurred. The oldest layer, at the bottom of the section, possibly shows a reversal transition direction.

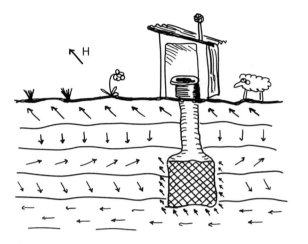

FIGURE A.2 This cartoon shows the record of remanent magnetization in various geological layers. The present direction of the magnetic field (represented by H) is up, as it would be in Australia. The upper layer formed recently and records the present direction of the magnetic field. Assuming the continent has not moved during the time of the formation of these layers, every successive layer in this figure records a reversal in Earth's magnetic field (see chap. 2), except for the bottom layer, which may reflect a transition state. The outhouse shows how liquids can lead to chemical alteration that resets the magnetization. The remanent magnetization near the base of the outhouse was acquired in the present magnetic field. The above cartoon by Charles Barton was redrawn by Beth Tully.

In practice, one would need the latitude of the site to arrive at precise interpretations.

Chemicals in the fluids emitted from a leaky outhouse in figure A.2 led to a new chemical remanent magnetization (CRM). A CRM is any remanent magnetization produced by a chemical change. The CRM in figure A.2 records the direction of the present magnetic field in the sediments in the vicinity of the outhouse. It could be detected from the magnetization's spatial distribution and from laboratory analyses showing that the magnetic minerals in the sediment nearest to the outhouse have different chemical compositions from minerals elsewhere in the sediment layers.

One common example of CRM involves the alteration of an existing ferromagnetic (or ferrimagnetic; chap. 1) mineral to another one. In some cases, a magnetic coupling occurs between the original and altered mineral phases that requires the new magnetization to be parallel to the magnetization in the initial mineral.[5] Therefore, the magnetic direction in the new mineral is identical to that of the original mineral. However, the intensity of the magnetization will have changed because the two minerals have different compositions: some of the atoms in these minerals have different atomic magnetic moments.

The measurement of magnetization before and after oxidation, a kind of chemical alteration, of magnetite at low temperatures under laboratory conditions indicates that this happens in practice. This result also explains why chemical change (oxidation) of oceanic basalts does not alter the magnetic stripe record, although the chemical change reduces the intensity of the magnetic anomalies (chap. 2). The intensity of the magnetic anomalies is found to gradually decrease with distance from the spreading ridge. (A greater distance implies a longer time since the basalts cooled; chaps. 2 and 6.)

While in the case just mentioned the CRM inherited the magnetic direction recorded when the rock formed, there are other types of CRM that record Earth's magnetic field when the chemical alteration occurs. This can happen, for example, when non-magnetic material is chemically altered to magnetic material. Because a CRM is a form of secondary magnetization that can be very stable, paleomagnetists carry out a number of experiments to sort out the magnetic history of the rocks they are studying. Generally speaking, rocks in which the magnetic minerals show no, or very little, chemical alteration are preferred to rocks that exhibit substantial amounts of chemical alteration.

Paleointensity

A magnetic field is represented by its direction and intensity. I have said only a little about paleointensities, estimates of the past intensity of a magnetic field. There are many paleointensity estimates for Earth, our Moon, meteorites, and even a few from Mars. The estimates for Mars come from meteorites recovered on Earth that originated on Mars when collisions with asteroids or meteorites sent off material into space, some of which found their way to Earth (chap. 3). Because Earth has a significant non-dipole magnetic field (chap. 1), scientists attempt to obtain an estimate of the intensity of the dipole field by using many measurements averaged over time.

Although there are different methods used to estimate paleointensities from lava flows, scientists often use one pioneered by the French husband and wife team of Émile and Odette Thellier, who published their initial results in 1937, well before magnetic field reversals were widely accepted. Following techniques described in classic papers by them in 1959, the Thellier method is still deemed the best we have, although many other methods are sometimes used. (For reasons I will not give, the Thellier method is not always applicable.) All intensity methods are based on observational evidence that igneous rocks typically acquire a thermal remanent magnetization (TRM) linearly proportional to the intensity of the magnetic field in which they cooled. One can measure the magnetization in a laboratory and multiply it by a constant to obtain an estimate of the ancient magnetic field strength, a paleointensity estimate. But what is the magnitude of the constant of proportionality? The Thelliers found this constant was different for every rock they measured. To determine it for a given rock sample, they heated the sample to temperatures above the Curie temperatures of all the magnetic minerals in the rock and then cooled it in a known magnetic field. Then, by measuring the new laboratory-produced TRM and knowing the intensity of the magnetic field, they could determine the constant of proportionality relating magnetization to field intensity. This then could be used to estimate the paleointensity.

A problem emerged. The constant of proportionality is affected by chemical change, which can occur during the experiment in a scientific laboratory or under natural conditions. The Thelliers found a way around this problem. They discovered that rocks did not lock in their TRM at a single temperature. Instead, rocks lock in their magnetization over a wide range of blocking temperatures. This can be appreciated by recalling Néel's theory for TRM. The relaxation time, which is used to define the blocking temperature, in the simplified version of Néel's theory depends on the shape of the grain. As there are typically many magnetic grains with different shapes in a sample of basalt, there are many blocking temperatures. The Thelliers found that the magnetization locked into the rock—say, between 580°C and 550°C—usually had a different constant of proportionality than that locked in between 550°C and 500°C.

The different blocking temperatures provided them with a way to develop a consistency check to determine if secondary magnetization (such as caused by chemical change) had wiped out reliable paleointensity information. The TRM acquired in any temperature interval—say, between 580°C and 550°C—in a laboratory can be compared to the initial TRM lost during heating in the same temperature interval to provide a paleointensity estimate.[6] This estimate can then be compared to another estimate obtained for an entirely different temperature interval, say, between

550°C and 500°C. The results should agree, within experimental uncertainty, providing that the primary magnetization was an unaltered TRM and the rock originally cooled fast enough that the magnetic field intensity did not have time to change. Such consistency checks were developed by the Thelliers and have been applied using their original version (which is more complicated than described here) or in various modified versions referred to as modified Thellier techniques.

There are methods besides the Thellier method for determining paleointensities. Although all methods employ consistency checks, problems have emerged to make some aspects of paleointensity studies very controversial. On rare occasions even the best consistency checks can fail, and the consistency checks employed by some non-Thellier methods are poor. Moreover, even if one can abstract a reliable paleointensity from a lava flow, it represents only an estimate at one location at one time. However, the intensity of the field can change relatively rapidly on a geological time scale, as evidenced by direct measurements made during the previous century (chap. 1). This produces a host of problems on how to average intensities from many flows to represent Earth's dipole magnetic field. Although I do not discuss these problems here, they are discussed in the reference given in note 6.

The most reliable paleointensity estimates occur when both absolute and relative paleointensity estimates provide consistent interpretations. Absolute paleointensity estimates refer to numerical values for the magnitude of the paleomagnetic field. These estimates are typically obtained from basalts using methods pioneered by the Thelliers and extended by first-class paleomagnetists, such as Rob Coe (University of California at Santa Cruz), Lisa Tauxe (University of California at San Diego), and Masuru Kono (Tokyo Institute of Technology). Relative paleointensity estimates, which do not provide numerical values, are typically obtained from continuously deposited sediments. For example, measurements from deep-sea sedimentary cores indicate that the paleointensity has been decreasing for about 2,000 years, a conclusion consistent with absolute intensity estimates obtained from basalts. However, a decrease in the intensity of the remanent magnetization with time in sediments is by itself insufficient to conclude that the magnetic field intensity also decreased: the mineralogy may have changed with time or some of the primary magnetization may have undergone alteration. Therefore, a host of rock magnetic analyses are undertaken before making relative paleointensity estimates to allow scientists to select sediments for which the source of the magnetic minerals remained constant and the primary magnetization unaltered. Nevertheless, errors still sometimes occur, leading to different interpretations of the intensity data.

Because of the problems mentioned above, some aspects of paleointensity research remain controversial. Therefore, I have not discussed paleointensities extensively in this book. I refer to paleointensities only when I believe the results are relatively uncontroversial or, in a few instances, when I indicate more measurements are required for confirmation. For example, paleomagnetists have little doubt that the intensity of Earth's dipole field decreases substantially during a magnetic field reversal (chap. 2). However, the magnitude of this decrease—for example, during the last magnetic field reversal—is not precisely known.

Notes

Chapter One

1. Following the lead of modern-day planetary scientists, I use "Earth" rather than "the earth" to describe our planet. This reflects the fact that we do not refer to planets such as Venus as "the venus."

2. Historians sometimes scold scientists for not taking the same care discussing history as they do science. I am guilty of such a charge. Although I have tried to minimize the number of references I give, I have tried to cite a reference when it is not primary and when I believe it may be controversial. There are many useful sources on the history of magnetism and related subjects. Here are a few books I particularly enjoyed reading: A. R. T. Jonkers, *Earth's Magnetism in the Age of Sail* (Baltimore: John Hopkins University Press, 2003); William Glen, *The Road to Jaramillo* (Stanford, CA: Stanford University Press, 1982); and James Newman, ed., *The World of Mathematics*, 4 vols. (New York: Simon and Schuster, 1956). Although one should be cautious about using material on the Internet, it can be valuable when you are familiar with the source. For example, I found articles by David Stern, a geomagnetist, to be useful. The reader should be cautioned that I often do not explicitly reference the above sources (and others), even though they have been used extensively throughout this book.

3. Such an experiment was actually done by Peregrinus and reported on in the *Epistola de magnete* in 1269, which he wrote while he was part of an army besieging the city of Lucera in Italy.

4. This dip occurs because I represented Earth's magnetic field with a dipole aligned along Earth's rotation axis with its north pole below (in the Southern Hemisphere) its south pole. This can be confusing because it often does not fit with a common preconception. The suspended magnet points downward in the Northern Hemisphere because the north pole of the magnet is attracted to the south pole of the magnet at Earth's center. This south pole is in the Northern Hemisphere. Although this sounds contradictory, it isn't. Physics requires the north end of the

suspended magnet to be attracted to the south end of the magnet at Earth's center. The confusion comes from the definition of the geomagnetic north pole, which is defined by the direction the north end of a compass needle points. The north end of a suspended magnet above the magnetic north pole would point straight downward because it is attracted to the south end of the "magnet" at Earth's center.

5. I have been a little loose in describing a metal paper clip as a paramagnetic material. The paper clip actually has a soft multidomain magnetism, which is described in the appendix.

6. A more lengthy discussion of how a physical scientist typically views magnetic therapy can be found in Robert Park, *Voodoo Science: The Road from Foolishness to Fraud* (Oxford: Oxford University Press, 2001). However, not stated in Park's book or in the body of my text, hemoglobin is paramagnetic when it is not carrying oxygen, that is, in the return of blood in veins to the heart. For chemists, the difference in the magnetic state of hemoglobin occurs because the iron ion is ferric in oxygen carrying blood and ferrous in non-oxygenated blood.

7. The exchange force is actually an electric force coupled with the so-called Pauli exclusion principle. This principle forbids two electrons from simultaneously occupying the same quantum state. Because of this principle, two electrons with the same spin directions are sometimes required to be farther apart than spins with the opposite directions. When this occurs, the electric force between the two electrons is smaller. The actual calculation of exchange is complicated because iron, a transition metal, has 26 electrons, each of which has a position that is described in terms of a probability function (the square of the electron's wave function).

8. As might be expected, NMR can only be explained using quantum mechanics. A strong magnetic field, tens of thousands times Earth's magnetic field, lines up the nuclear dipole magnets. They are then perturbed by applying an alternating magnetic field in the radio frequency range. This field is applied perpendicular to the original field. The magnetic dipole moments then deviate from the direction of the strong magnetic field. They also precess (revolve around) the direction of the constant field. Energy from the radio frequency field can be absorbed under situations I will not describe, but require the frequency to be close to that of the precessing nuclear dipoles.

9. Joseph Needham, *Science and Civilization in China*, vol. 4, *Physics and Technology, Part 1, Physics* (Cambridge: Cambridge University Press, 1962). Needham also attributes the invention of the first (land) compass to the Chinese over 2,000 years ago. However, not all historians credit the Chinese with these discoveries.

10. Remagnetization was achieved by heating the needle above its Curie temperature and cooling it while the needle faced in a northerly direction, as determined by the stars.

11. According to the historian William Glen, *The Road to Jarmillo* (Stanford, CA: Stanford University Press, 1982), the difference between true and magnetic north was not known in Europe until the sixteenth century, even though the Chi-

nese had learned about this difference by 720 AD. The English historian A. R. T. Jonkers wrote an excellent account of magnetism and ancient mariners: *Earth's Magnetism in the Age of Sail* (Baltimore: Johns Hopkins University Press, 2003). Many of the historical anecdotes used in this chapter come from his book.

12. Vector addition must be used. It should also be pointed out that the dipole field and nondipole field cannot be distinguished using only data from a single point on Earth's surface.

13. The quadrupole example (fig. 1.4) is called an axial quadrupole field. There are other types of quadrupole fields, as described later in n. 25.

14. For example, the variation in temperature with time in ice cores can be estimated from the ratio of oxygen-18 to oxygen-16, where the number following "oxygen" refers to the number of neutrons and protons in the nucleus. The heavier isotope, oxygen-18, condenses out of the atmosphere faster when the temperature falls. This allows one to use the oxygen isotope ratio to estimate temperatures in samples from ice cores: the higher the ratio of oxygen-18 to oxygen-16, the lower the temperature.

15. Milankovitch proposed in 1941 that variations in solar radiation occurred because of changes in the orientation of Earth's rotation axis and because of changes in Earth's orbit. Precession of the Earth's rotation axis occurs near 23,000 years, Earth's rotation axis changes its tilt with a periodicity of 41,000 years, and the eccentricity of Earth's orbit changes with a 100,000-year periodicity. The "eccentricity" refers to the ratio of the primary (longest) axis to the secondary (shortest) axis of the elliptical orbit of Earth around the Sun. As an aside, Earth's elliptical orbit deviates only slightly from a circular one.

16. Direct radiation effects associated with the orbital changes given in note 15 seem too small to produce the effects recorded in ice cores. This means that some amplification process is required. Speculations as to the nature of this amplification process have been made as well as speculations on why there was a shift from a 41,000- to a 100,000-year period about 900,000 years ago. However, there is no widespread agreement concerning the validity of these speculations.

17. The probability level used for this is somewhat arbitrary; it varies for different scientific fields. A 95 percent probability that some result did not occur by chance is commonly used in analyses of problems in Earth magnetism.

18. This is an example of an authority argument that can be checked. Flip a coin a hundred times and record the results. If you repeat this many times, you can accurately evaluate my assertions. However, even if you do this only once or twice, I think you will conclude I am probably correct.

19. This quote comes from part of the title of a paper presented by Edward Lorentz at an American Association for the Advancement of Science meeting held in Washington, D.C., in 1972.

20. Some mathematicians point out that sensitivity to initial conditions is not a necessary requirement of deterministic chaos, contrary to many statements made

in the popular and scientific literature. Even when such sensitivity is present in one space, one can sometimes change the metric (the way distance is measured) to remove this sensitivity. Although some mathematicians may complain, I have chosen to ignore this objection to produce insight into deterministic chaos for someone not wanting a complete technical explanation, which would involve a lengthy discussion of nonlinear dynamics.

21. Adrien-Marie Legendre, another famous mathematician, also receives credit for this, having developed it independently in 1806.

22. Eric Bell, "The Prince of Mathematicians: Gauss," appearing in Vol. 1, *The World of Mathematics*, ed. James Newman (New York: Simon and Schuster, 1956), p. 295.

23. This includes one of my own co-written books: R. T. Merrill, M. W. McElhinny, and P. L. McFadden, *The Magnetic Field of the Earth: Paleomagnetism, the Core, and the Deep Mantle*, (San Diego: Academic Press, 1998).

24. Halley used Barlow's 1833 chart for declination, Horner's 1836 chart for inclination, and Sabine's 1837 chart for total intensity. As these come from different years, Gauss's description would not constitute an International Geomagnetic Reference Field (IGRF) for a particular year (defined later in the text). The best reference fields were not produced until the twentieth century. Reference fields established from ancient mariners' records had to make an assumption concerning intensity, because intensity data were not available until the nineteenth century. It turns out that all one needs is one good estimate of the average dipole intensity to determine an IGRF for that year.

25. In a spherical harmonic expansion, there are three terms required to describe a dipole field and five for a quadrupole field. The next higher-degree field is referred to as the octapole field, which requires seven terms to describe it fully in a spherical harmonic expansion. While the dipole field falls off inversely as the cube of the distance from the source, the quadrupole field falls off as the fourth power of the distance, the octapole field as the fifth power of the distance, and so on. The higher the degree of the harmonic, the more terms that are required to describe it and the faster the field dies off with distance from the source. Similar to reasons why the dipole field decreases faster with distance than the monopole one, higher-degree nondipole field terms fall off faster with distance from the source than do lower-degree ones.

26. A decline in the dipole field does not require the nondipole field to increase to conserve energy. This can be illustrated by considering the magnetic field near a current loop driven by a battery. Such a field will have both dipole and nondipole components. When the battery dies, both the dipole and nondipole field decay away with time. The magnetic energy lost goes into heat, a different form of energy. In the case of Earth, the energy is primarily partitioned between the magnetic field, heat, and fluid motions.

27. The Pacific Ocean does not affect the nondipole field. This description is used to provide a geographical reference. It is not clear, and there are conflicting speculations, why the nondipole field is relatively low in the Pacific hemisphere.

28. Some readers may have been taught in an elementary physics course that a circular electric current loop produces a dipole magnetic field. This is not strictly true close to the loop, where both dipole and nondipole fields are present. Because nondipole fields rapidly die off with distance from the loop, most elementary physics books do not describe them. The magnetic field far from a circular current loop appears as though it were coming from a dipole magnet at the center of the loop and perpendicular to the plane of the loop. Other complications, not given here, are discussed in the reference given in note 23.

Chapter Two

1. Although the minerals used to identify granite cannot carry a remanent magnetization, granite often contains minor amounts of other minerals, a few of which can be stably magnetized. This is further discussed in the appendix.

2. Remanent magnetization is the magnetization a sample has when no magnetic field is present, besides that produced by the sample itself. An induced magnetization occurs in the presence of a magnetic field outside the sample, such as Earth's magnetic field. When this field is reduced to zero, the induced magnetization vanishes.

3. Several minerals, including iron oxides and sulfides, are capable of carrying a remanent magnetization. However, for simplicity's sake, I will use magnetite to illustrate general principles. Similarly, many igneous and sedimentary rocks are used in paleomagnetic studies, even though I emphasize basalt.

4. Although magnetite in most terrestrial basalts contains only small amounts of titanium, large amounts of titanium are common in magnetite, more properly called titanomagnetite, in marine basalts. The amount of titanium present in magnetite strongly affects its Curie temperature. A typical Curie temperature for a titanomagnetite in a submarine basalt is only around 150°C (302°F). In contrast, iron, which is the common recorder of magnetization on our Moon, has a Curie temperature of 770°C (1,418°F).

5. For example, the following quote comes from Sir Harold Jeffreys, *The Earth: Its Origin, History, and Physical Constitution*, 4th ed. (Cambridge: Cambridge University Press, 1959), p. 371: "When I last did a magnetic experiment we were warned against handling permanent magnets, as the magnetism was liable to change without much carelessness. In studying the magnetism of rocks the specimen has to be broken off with a geological hammer and carried to the laboratory. It is supposed that in this process its magnetization does not change to any important

extent, and though I have often asked how this comes to be the case I have never received an answer."

6. A history of this time has been published by one of the participants, E. Irving, "The Role of Latitude in Mobilism Debates," *Proceedings of the National Academy of Sciences* 102, no. 6 (2005): 1821. Edward Bullard used a computer in 1965 to show that the fit of the Atlantic continents at the 500-fathom contour line was excellent, a support of Wegener's conjecture.

7. The geocentric axial dipole assumption is viewed today as a very good (but not perfect) assumption. However, a longer time interval than 10,000 years is preferred. A minority of geomagnetists advocate that the assumption fails for some geological intervals that ended more than 300 million years ago. There are many recent reviews on this subject, including one of ours: R. Merrill and P. McFadden, "The Geomagnetic Axial Dipole Field Assumption," *Physics of the Earth and Planetary Interiors* 139 (2003): 171.

8. In practice, scientists calculate a pole using a magnetic direction from a single rock unit, say, a lava flow. Because a lava flow cools far too quickly to average out secular variation, this pole is called a virtual geomagnetic pole, or VGP. A VGP usually does not represent an accurate location of the ancient north pole. Instead, many VGPs are averaged to obtain an estimate of the paleomagnetic pole.

9. One degree of arc distance is equal to 111 km (69 miles) at Earth's surface.

10. E. Irving, *Paleomagnetism and Its Applications to Geological and Geophysical Problems* (New York: John Wiley, 1964).

11. European universities commonly award a doctorate of science degree, called a science doctorate degree (ScD), while American universities do not. It is usually awarded when a university concludes that one of its ex-students has made a particularly important contribution to science. It is a more prestigious degree than a PhD degree.

12. These lavas sometimes heated up the upper part of sediments upon which they flowed to sufficiently high temperatures that, when the sediments cooled, they also acquired a new magnetization parallel to Earth's magnetic field. This sediment magnetization, recorded in baked clays, provided the first so-called field check. It indicated that the magnetizations in lavas and clays, which were acquired at the same time, recorded the same magnetic field directions. If this had not been the case, Brunhes and David would have concluded that the magnetic record in either the lavas or their baked clay contacts had changed after the rocks had cooled: the fossil magnetic record would not have been trustworthy.

13. William Glen wrote an excellent account of the history of this subject: *The Road to Jaramillo* (Stanford, CA: Stanford University Press, 1982).

14. The situation has changed since the time of Blackett. Now scientists are aware of rapidly rotating stars that have only weak magnetic fields. A rapidly rotating star (which has more than 1.5 the mass of our Sun) can rotate at a rate less than one Earth day (more than 25 times that of our Sun). The outer part of such a star

lacks the convection zone needed to produce a dynamo (chaps. 3 and 4). Rapidly rotating stars "spin up" quickly because they form by collecting material from a rotating molecular cloud (chap. 3). This is similar to an ice skater increasing her spin rate by moving her arms inward. A moderately strong magnetic field is needed to shed angular momentum to counter this. The observation of such "fast rotators" shows there is no simple correlation between faster rotation and magnetic field strength, as was once erroneously believed.

15. Néel received the Nobel Prize for discovering antiferromagnetism. Iron is ferromagnetic, meaning adjacent atomic magnetic moments (like very tiny magnets) line up parallel to one another. In antiferromagnetic substances, such as ilmenite, the adjacent atomic magnetic moments line up antiparallel to each other (see fig. 1.2). Although I use "Curie temperature" throughout this book, technically one should use "Néel temperature" whenever antiferromagnetic coupling is involved. John Graham also asked Néel in a letter whether self-reversal of magnetization was possible.

16. Néel suggested a few different mechanisms for self-reversal. All of them involve the magnetic field interaction between two different minerals. The first mineral has a higher Curie temperature and acquires a magnetization parallel to the external field. The second mineral acquires its magnetization later during cooling because it has a lower Curie temperature. Néel showed that in some cases the second mineral could be affected strongly enough during cooling by the magnetic field of the first one to become magnetized in the opposite direction. For such mechanisms to work, the first magnetic mineral has to have its magnetization locked in at a higher temperature than the second one, and the magnetization associated with the second mineral has to dominate the magnetization of the first one at room temperature.

17. For the reader well versed in the physics of magnetism, the Haruna Dacite contains two different phases of titanohematite that exhibit a negative exchange interaction with each other. The details of this mechanism are still being debated in today's scientific literature.

18. Uyeda describes his work and beliefs on this subject in his book *The New View of the Earth: Moving Continents and Moving Oceans* (San Francisco: W. H. Freeman, 1978).

19. I was a student at Berkeley from 1961 to 1967, and John Verhoogen was my PhD adviser. The historical account given here differs only slightly from that of Glen (n. 13), who wrote his book after interviewing the major participants in the reversal controversy.

20. Potassium-39 and potassium-41 are stable isotopes that have 20 neutrons and 22 neutrons respectively in their nuclei. Unlike potassium-40, they do not undergo radioactive decay.

21. It did not take long for geologists to realize this was oversimplified: certain minerals, such as hornblende, retain argon better than others, such as biotite.

Geologists found they had to use minerals that had not lost argon through chemical alteration. Much later it was discovered that not all the gas was lost when oceanic basalts erupted far below the sea surface. The high pressure of the water above the flows inhibits the complete outgassing of argon. A new variant of this method evolved during the 1980s that uses different isotopes of argon. This method is now preferred to the method described in this chapter. The time scales used by geologists continue to be refined as old methods of dating evolve and new methods are introduced.

22. Assuming the two buckets are the same size, the red bucket must have been 1/2 full at the start. Fifteen minutes later it would have been 1/4 full, and 15 minutes after that it would have been 1/8 full. So the answer is 30 minutes.

23. G. Brent Dalrymple, *The Age of the Earth* (Stanford, CA: Stanford University Press, 1991). Also see G. Brent Dalrymple, *Ancient Earth, Ancient Skies: The Age of Earth and Cosmic Surroundings* (Stanford, CA: Stanford University Press, 2004). The oldest rocks on Earth have been dated to be about 4 billion years using uranium isotopes. However, some radioactive uranium is found in minerals (zircons) in rocks from western Australia that yield dates of 4.3 to 4.4 billion years. These minerals are found in younger sediments, which in turn had to have formed after older rocks were eroded. Minerals from these older rocks, including zircons, were then deposited as sediments. The older rocks are no longer present, having been assimilated back into Earth's mantle.

24. For example, see T. Barnes, *Origin and Destiny of the Earth's Magnetic Field* (San Diego: Creation-Life Publishers, 1973). More recently, Russell Humphreys has extended the creationists' arguments.

25. "Peer review" usually means two or more reviewers, judged to be experts in the field, will be sent the manuscript and asked for comments. These reviewers may recommend against publication. However, even if they recommend that the manuscript be published, it is rare when they do so without also requesting some alterations. This typically leads to manuscript revisions. Ultimately, the editor of the journal makes the final decision to publish or not. Although exceptions occur, it often takes a year or more from the first submission date of a manuscript for it to appear in a journal. (A few journals, such as *Nature* and *Science*, are exceptions.) Even though this process requires time and effort on the part of many scientists, it is a valuable one because it provides some necessary checks and balances on what is published. In particular, it helps weed out pseudoscience articles from being published in respectable scientific journals.

26. The first grant I ever received was from the Office of Naval Research to study the origin of a magnetic anomaly of a seamount off the coast of Washington state.

27. Glen, *The Road to Jaramillo*, p. 299.

28. L. Morley and A. Larochelle, "Paleomagnetism as a Means of Dating Geo-

logical Events," in *Geochronolgy in Canada*, ed. F. F. Osborne (Toronto: University of Toronto Press, 1964), pp. 39–51.

29. The first reversal event was discovered in rocks from the Olduvai Gorge in Tanzania by University of California at Berkeley scientists Sherman Grommé and Dick Hay.

30. This is contrary to what you might read in some popular books and on the Internet. The absence of any reversal during the past 780,000 years has been generally accepted by geomagnetists for the past two decades or so, but not before that.

31. Reversal intervals were originally named after famous geomagnetists, as shown in figure 2.1. Although a newer technical terminology for reversal intervals has now been adopted, neither is used in this book.

32. Geologists divide Earth's history into two super-eons. The Precambrian super-eon extends from the birth of Earth 4.55 billion years ago to 544 million years BP when a major increase in biological diversity, including the widespread use of shells in many animals, occurred to usher in the Phanerozoic super-eon.

33. A few papers have been published in peer-reviewed journals claiming the existence of subtle periodicities in the magnetic field reversal chronology. Although a couple of these claims have not been convincingly refuted, they have also not been convincingly demonstrated. If there is any periodicity, it is very slight and embedded in a record dominated by stochastic processes.

34. From *The Knowledge Book* (Washington, DC: National Geographic Society, 2007), p. 150.

35. This excludes the rare self-reversing rock.

36. A few older superchrons have been claimed to exist, but they have not been widely accepted because the reversal chronology for older times is not yet reliable.

37. To my knowledge, the first suggestion of a superchron came from Ted Irving and L. Parry in 1963.

38. Some scientists have suggested one or more short reversal events may have occurred during the two best documented superchrons. Although one cannot rule out this possibility, based on theoretical reasons it is unlikely that many events have been missed. Theory indicates that the average reverse and normal polarity magnetic fields are symmetric, other than the difference of the sign of the magnetic field. This is consistent with all modern statistical analyses of the reversal chronology data. Neither the normal nor the reverse polarity state appears preferred over the other. If you flipped over the average reverse magnetic field, you could not distinguish it from the normal polarity field. It would exhibit the same average directions and intensities. If there had been more than a few short reverse states in the long normal polarity interval that began 118 million years ago, this would have broken the symmetry. It would indicate that Earth had a bias toward the normal polarity state, contrary to theory. Of course, the theory could be wrong, and if so, this would imply some curious phenomena occurred. In any case, Earth is still divided

into two mega-states: one in which reversals are relatively common and one in which reversals don't occur or infrequently occur.

39. Creationists particularly like this result, because this means they can accept that magnetic field reversals occur without requiring Earth to be very old. You cannot have many field reversals, each taking more than a thousand years, and still claim Earth is only about 6,000 years old. However, you can have many reversals and have a young age for Earth if each reversal occurs in a time interval of less than a year.

40. For statisticians, Phil McFadden and I use a gamma distribution in our analyses of polarity interval lengths. The details of the statistics of reversals are reviewed in our book: R. T. Merrill, M. W. McElhinny, and P. L. McFadden, *The Magnetic Field of the Earth: Paleomagnetism, the Core, and the Deep Mantle* (San Diego: Academic Press, 1998).

41. While aficionados of chaos theory often claim that reversals of Earth's magnetic field illustrates deterministic chaos in action, no one has shown, or not shown, this to be the case. Instead, grossly oversimplified dynamo models are used to show that deterministic chaos *might* be occurring.

42. Although I liked and respected Ling, the joint work never occurred because I was then interested in the problem of how rocks recorded and stored information on Earth's magnetic field; I was not interested in using this record to reconstruct geological history.

43. This extinction, which occurred 65.5 million years ago, is commonly called the K-T extinction. K stands for the Cretaceous period and T stands for the Tertiary period, a term now replaced by the Paleogene and Neogene periods. Although most people identify the K-T (or the Cretaceous-Paleogene) boundary as the time when dinosaurs went extinct, paleontologists find that fossils of smaller and more numerous animals are easier to use to identify this boundary. For example, because dinosaur fossils are uncommon, they are more difficult to use to define the precise location of the K-T boundary.

44. Scientists, such as Chuck Officer, point out that some volcanic eruptions also produce large amounts of iridium. Officer advocated that the mass extinctions occurring 65 million years ago were caused by the massive outpouring of basalts in India know as the Deccan Traps. Later scientists, such as Vincent Courtillot, extended these arguments (chap. 6).

Chapter Three

1. I occasionally use information (usually unacknowledged) from papers of the science historians Stephen Brush and Mott Green. For example, in this section I have made use of Brush's paper "Inside the Earth," *Natural History* 93 (1984). The earliest seismometer (as opposed to a seismoscope, which does not record elastic

waves) that I am aware of came from the team of John Milne, James Ewing, and Thomas Grey, who worked in the 1880s in Japan.

2. Joseph Needham, *Science and Civilisation in China*, vol. 3, *Mathematics and the Sciences of the Heavens and Earth* (Cambridge: Cambridge University Press, 1959), pp. 624–35.

3. Modern seismometers use horizontal or inverted pendulums and rely heavily on various electronic devices.

4. The core is best recognized from the presence of a "shadow zone," a region on Earth's surface where P waves do not arrive. This shadow zone extends from an arc distance of 105° to 143° from the source. (Seismologists often use arc distances in their work. One degree of a great circle arc on the surface of Earth corresponds to 111 km, or 60 miles.) The shadow zone exists because of refraction, which occurs across interfaces that exhibit jumps in wave speed. Consider a pencil partially extended into a glass of water. The pencil appears to bend at the point it enters the water. This optical illusion occurs because the speed of light in water is lower than that in air: the direction light travels abruptly changes across the water-air interface. The speed of P waves near the bottom of the mantle is slightly greater than 13 km per second and it drops to around 8 km per second as the P waves enter the core. (The 8 km per second speed at the top of the core is the same speed P waves have at the top of the mantle.) Because of this change in wave speed, refraction occurs at the core-mantle boundary causing a P wave to bend toward the vertical (radius) as it enters the core. It again refracts (away from the vertical) on exiting the core. This refraction results in the absence of P wave arrivals in the arc distance range between 105° and 143° from the epicenter. This is the shadow zone. These and other data, including those dealing with reflections of seismic waves off the core-mantle boundary, have produced a clear picture of the core.

5. Earth is slowing down due to tidal interaction with the Moon. Our Moon moves away from us to conserve angular momentum. Laser light bounced off a mirror placed on the Moon during the Apollo missions accurately determined the change in distance over time between Earth and our Moon. The data indicate that the Moon is retreating from us at a rate of 3.8 cm (about 1.5 inches) per year.

6. The details of this collapse are uncertain and model dependent. For example, one scenario argues that solid material first formed to produce planetesimals (small planet-like objects), which accreted into planet cores in the case of the giant planets, Jupiter, Saturn, Neptune, and Uranus. The core's gravitational field subsequently accreted the gases that make up the bulk of the outer planets. The second scenario involves fluctuations within the gas disk itself, which produces regions with increased density that serve as centers for planet accretion. The main testable difference between these two scenarios is that the first scenario predicts larger cores for the giant planets. This test has not yet been carried out.

7. As quoted in G. Brown and A. Mussett, *The Inaccessible Earth*, 2nd ed., (London: Chapman and Hall, 1993), p. 61.

8. More properly I am describing a subclass of carbonaceous chondrites referred to as C1 chondrites.

9. The fusion in the Sun begins with the combination of four hydrogen atoms to form a helium nucleus. (Two electrons are lost in this process. Because the Sun consists of plasma, i.e., ionized gas, "hydrogen" consists of one proton and one electron that are not bound to each other.) A helium nucleus contains two neutrons as well as two protons. The hydrogen isotope deuterium contains one neutron. The helium isotope (helium-3) also contains one neutron. These isotopes are also made in the Sun but in small amounts.

10. Although the composition of the Sun, meteorites, Earth, and other planets all originated when the solar system formed, the inner planets (Mercury, Venus, Earth, and Mars) did not have strong enough gravitational fields to hold on to much of the gaseous material that make up the bulk of the outer planets (Jupiter, Saturn, Neptune, and Uranus). Also see note 6.

11. The age of the Sun can be estimated in several ways. One way is to observe many stars that are similar to our Sun and see how they evolve with time. This evolution considers nuclear synthesis theories that explain how chemical composition changes with time. For example, the fusion of hydrogen to helium cannot go on forever because the Sun would run out of hydrogen.

12. F. Birch, "Elasticity of the Earth's Interior," *Journal of Geophysical Research* 57 (1952): 227–86.

13. Initially Ringwood used analogue minerals (germanates), because he could not obtain sufficiently high pressures to produce phase changes in olivine and pyroxene.

14. A series of solid-solid phase changes convert olivine and pyroxene to more dense minerals between a mantle depth of 410 and 670 km. The region bounded by the 410 and 670 km transitions is referred to as the mantle transition zone. The most common mineral in Earth is the magnesium-iron-rich perovskite, which occurs throughout most of the mantle below a depth of 670 km. Its composition is estimated to be $Fe_1Mg_9SiO_3$. This perovskite was discovered by Lin-gun (John) Liu, a colleague of Ringwood's at the Australian National University.

15. A cautionary note to historians: although I knew Ringwood well, I never asked him about this incident. On the two occasions I asked Crawford, he simply smiled and changed the subject. The account given in the text comes from other faculty members at ANU who told me about the incident.

16. Different aspects of the Kelvin paradox are discussed in the following two articles: Frank Stacey, "Kelvin's Age of the Earth Paradox Revisited," *Journal of Geophysical Research* 105, no. 13 (2000): 155–58; and P. England, P. Molnar, and F. Richter, "John Perry's Neglected Critique of Kelvin's Age for the Earth: A Missed Opportunity in Geodynamics," *Geological Society of America Today* 17 (2007): 4–7.

17. Holmes's estimate was off the mark for several reasons, including the incor-

rect assumption that the rocks he measured contained no initial lead: all the lead was assumed to have formed by the decay of uranium.

18. The sizes of ocean basins and continents are changing with time. Between about 300 and 175 million years ago, only one supercontinent, Pangaea, and one super-oceanic basin existed. Darwin's fission hypothesis also does not explain the Moon's small iron-rich core. In addition, numerical calculations (not discussed in this book) involving conservation of angular momentum and energy in the Earth-Moon system are inconsistent with this model.

19. Bill Hartmann and Donald Davis first proposed this model in 1975. Shortly thereafter, Al Cameron and William Ward published a (then-) sophisticated computer model suggesting that a Mars-sized body collided with Earth to produce the Earth-Moon system. The discussion in the text uses a model by Robin Canup, who published the main results of her PhD dissertation on this subject in 1997. The size of Theia, the impacting planet, is determined from the need to conserve angular momentum and energy in the system. The Moon formed much closer to Earth, arguably around ten Earth radii, and has been moving slowly away ever since because of tidal interactions.

20. The transition between water and ice is probably the most recognizable solid-liquid phase transition.

21. There can be some exceptions, such as the water-ice transition, in which water is denser than ice at the melting temperature. I ignore these exceptions, which require a detailed discussion of the nature of atomic bonding. For that matter, there are even different definitions of what constitutes a liquid. For example, the usual physics definition of a liquid is that it manifests only short-range order of its atoms. In this case, a solid transforms to a liquid when the long-range ordering of atoms breaks down to the short-range ordering of a liquid. This definition requires that glass, which only exhibits short-range order, is a liquid. In this book, I use the definition that a liquid does not transmit seismic shear (S) waves; therefore, glass is considered a solid.

22. While the inner-outer core boundary does represent a phase change, the core is not made of pure iron. The actual phase transition dividing the inner and outer core is more complex than that of pure iron. It likely involves a transition in which solid and liquid coexist, because the melting temperature (the liquidus) lies above the freezing temperature (the solidus) for materials that are thought to constitute the core. The weight of the overlying Earth may squeeze out most of the liquid in the transition zone so that the transition from liquid to solid is sharp: P wave reflection data have been interpreted as showing the transition is less than 1 km in thickness.

23. The convection in Earth's core is thought to be driven by two processes. The first is the temperature gradient between the outer and inner core. The second is more complicated and is described as chemical convection. When the inner

core freezes to almost pure iron, it releases material less dense than iron. This lighter liquid material is buoyant and rises to stir the outer core. This last process is thought to be the primary source of convection in Earth's core.

24. Because European scholars regarded the north end of a compass as being the most fundamental one and because its north end points to the pole star (Polaris, or the North Star), some early European philosophers, such as Jacques de Vitry in 1218 and Girolamo Cardano in 1550, concluded that the lodestone, with which the compass needle was rubbed, obtained its "virtue" from the pole star. That is, magnetic materials on Earth acquire their magnetism from the heavens.

25. I have had considerable difficulty tracing down primary references to Einstein's statements on Earth's magnetic field. I would appreciate receiving any primary reference the reader can find. The reference to Einstein used in the text comes from a U.S. National Academy of Sciences memoir on Walter Elsasser written by Harry Reid, which was available on the Internet in 2007.

26. As far as I know, the oldest rock carrying a primary magnetization has been dated at 3.5 billion years. This rock has passed the so-called fold test. Around 3 billion years ago, the rock was folded by tectonic processes, which resulted in the rock's remanent magnetization pointing in different directions on the two different sides of the fold. However, if the rock is unfolded on a computer, the remanent magnetization throughout the rock formation points in the same direction. This provides paleomagnetists with confidence that the rock acquired its magnetization before 3 billion years ago and likely when the rock formed. Theorists presume that Earth's magnetic field was produced in the core shortly after Earth was born, but this has not yet been demonstrated beyond a reasonable doubt.

27. Scientists refer to a relaxation time (rather than a half-life, as I used earlier) to describe this exponential decay. The field would decay by $1/e$ of its initial amplitude in about 15,000 years (e is an irrational number that is approximately equal to 2.71828). This decay time could be a third to three times the value given.

28. These mathematicians, who worked independently, showed that an infinite number of dynamos can exist in an infinitely sized convecting electrical conductor. This may seem abstract to the reader, but it provided an important incentive to search for a viable dynamo model in finite-sized systems. Although these mathematicians showed that dynamos exist, they did not find one that worked.

29. For specialists, the Maxwell equations indicate that the electric current (density) can be determined from the curl of the magnetic field.

30. The analogy suffers from the fact that heat is a scalar while the magnetic field is a vector.

31. This diffusion is found to depend inversely on the electrical conductivity. If the material is a good electrical conductor, as is the core, diffusion occurs slowly. If the electrical conductivity is perfect, no diffusion occurs and the magnetic field is frozen into the fluid. The field has a "free decay" of 15,000 years; it decreases by 63 percent in every 15,000-year interval if the fluid velocity is zero. No diffusion

occurs only in the limit of infinite conductivity. In practice, both diffusion and new magnetic fields are created in Earth's core when shear in the fluid flow occurs.

32. Technically, a perfect conductor differs from a superconductor. No magnetic field lines can exist in the so-called Type 1 superconductors. Explanations for a perfect conductor rely on classical physics, while those for superconductors rely on quantum mechanics.

33. For the scientific linguist, I should say that the heat is transferred by advection (rather than convection) in the outer core. If you put dye into a pan of boiling water, it is carried along (advected) with the water. In a similar way, heat is carried along (advected) in a convecting fluid. Also see note 23.

34. David Gubbins, "Geomagnetic Reversals," *Nature* 452 (2008): 165–67.

35. Although it is difficult to estimate accurately the velocity of the fluid in the outer core without making some controversial assumptions, reasonable estimates can be made from analyzing secular variation data. This subject is reviewed in R. T. Merrill, M. W. McElhinny, and P. L. McFadden, *The Magnetic Field of the Earth: Paleomagnetism, the Core, and the Deep Mantle* (San Diego: Academic Press, 1998).

36. Dave Stevenson at Caltech suggested that the slow rotation of Venus is not the reason Venus does not have a dynamo. He argues that the energy source in Venus's core is insufficient to sustain a convection-driven dynamo.

37. The strength of a magnetic field varies with distance from the source. To compare relative strengths of the magnetic fields, it is common to refer to the strength of the magnetic dipole moment of the planet. Jupiter's dipole moment is 2×10^4 M (where M stands for Earth's dipole moment, near 8×10^{22} Am^2), Saturn's is 5.9×10^2 M, Uranus's is 47.5 M, and Neptune's is 25 M.

38. Uranus has a dipole tilt of 59° with respect to its rotation axis, and the corresponding tilt for Neptune is 47°. However, the best fitting dipole is far from the center of each planet. This is another way of saying these planets have large nondipole fields (see chap. 1). A large dipole tilt does not necessarily imply a reversal is occurring.

39. A very strong magnetic field can distort the shape of a material. In agreement with the analogy between a rubber band and a magnetic field line, a magnetic field exhibits tension when stretched. Similarly, magnetic fields also exert an outward pressure. Magnetars are speculated to have crusts (unlike our Sun). The stress of a magnetic field moving through this crust has been speculated to produce starquakes, which release a large amount of energy in the form of elementary particles (electrons and positrons) and gamma rays.

40. Although these conditions are consistent with theory and observations, exceptions cannot be ruled out. For example, some theory suggests that conditions may exist by which a non-rotating planet could produce a dynamo.

41. Subsequent data confirm that Mars exhibits magnetic "stripes," but they are not as striking as reported in 1999. They look like adjacent elongated structures

that are oppositely magnetized. Planetary scientists still think a dynamo operated in Mars prior to 4 billion years ago and that it reversed polarity on occasion. The question of what caused tectonics on Mars is not settled. However, planetary scientists recognize any viable global tectonic model for Mars needs to explain the 4 km higher average elevation of the southern hemisphere (which covers 42 percent of Mar's surface) compared with the smoother northern hemisphere.

Chapter Four

1. The Doppler shift, named after Austrian astronomer Christian Doppler (1803–1853), is used to detect these motions. When the surface of the Sun moves toward us, the wavelength of the light emitted appears to become shorter (in the visible range, it would become bluer), and it appears to lengthen (redden) when the surface moves away from us. The Doppler effect is employed by police using radar to measure the speed of moving vehicles.

2. Although the solar system outside the Sun contains less than 1 percent of the solar system mass, it accounts for 99 percent of its angular momentum (the momentum of rotation).

3. Thomas Harriot, Johannes and David Fabricius, and Christopher Scheiner are also sometimes credited with using the newly invented telescope to observe sunspots at the same time Galileo did.

4. A short review of Newton's strong religious beliefs was written by Martin Gardner in *Skeptical Inquirer*, September 1996.

5. Indirect methods, such as the Zeeman effect, are used to estimate the magnetic field of stars, including our Sun. The Zeeman effect is based on quantum mechanics: electrons orbiting an atom can only emit light with certain discrete wavelengths. It was first discovered in 1896 by the Dutch physicist Pieter Zeeman, who received a Nobel Prize in 1902 for his discovery. Zeeman thought he observed a broadening of a spectral line for yellow light when he put sodium in a flame held between the poles of a strong magnet. In fact, what he discovered, as was confirmed later, is light emitted from sodium gas splits into distinct wavelengths (they are quanititized) when a magnetic field is present. These different wavelengths correspond to different shades of yellow, which is the color of the light given off by sodium. The magnitude of the splitting is proportional to the intensity of the magnetic field. This last observation allows astronomers to obtain magnetic field intensity data for our Sun and other stars.

6. The 22-year magnetic polarity cycle is called Hale's law, named after George Hale (1868–1938), who discovered the 22-year cycle in 1908. Horace Babcock qualitatively explained the Hale cycle in 1961, a cycle still imperfectly explained by modern mathematical dynamo models.

7. John Eddy produced several papers documenting the Maunder minimum.

Minze Stuiver at the University of Washington used carbon-14 measurements to support Eddy's contention that isotopes could be used to test if the minimum was real. As we shall discuss further in chapter 6, cosmic rays entering our atmosphere show an 11-year periodicity related to the sunspot cycle. More carbon-14 is produced when the number of cosmic rays entering our atmosphere increases. Eddy also showed that fewer auroras were observed during the Maunder minimum, consistent with observations that greater solar activity is correlated with a larger number of auroras.

8. The fluid motion exhibits large variations throughout the convection zone, which is about 30 percent (by radius) of the Sun. By contrast, the underlying radiation zone exhibits a remarkably uniform rotation. This provides valuable information for the theorist. Magnetic fields with radial components cannot extend far into the radiation zone. Otherwise they would disrupt the uniform (within the error limits of the inversion) flow of plasma inferred from helioseismology. The transitional layer between the convection zone and the radiation zones, the tachocline, has a thickness of about 5 percent or so of a solar radius. It plays a crucial role in modern solar dynamo theory. Modern-day solar dynamo models differ substantially from Earth dynamo models in the sizes and shapes of the convection cells. For example, the fluid moving downward in the Sun is usually confined to relatively smaller cylinder-shaped regions called plumes, while the upwelling regions are much broader. In addition, local dynamo action probably plays a crucial role in producing the magnetic fields of sunspots, but in a different way from that suggested by Larmor and opposed by Cowling (chap. 3).

9. For example, a different mechanism involves electromagnetic waves, called Alfvén waves (named after their discoverer Hannes Alfvén). These waves, which travel at speeds proportional to the square of the magnetic field, are hypothesized to carry energy from the Sun's surface to the corona, where they turn into shock waves and dissipate into heat.

10. The solar wind's supersonic flow ends at the so-called terminal shock boundary, beyond which the wind flows at a substantially reduced rate. The terminal shock boundary lies inward from the heliopause. *Voyager 2* found the terminal shock to be about 1.6 billion km closer to the Sun than *Voyager 1*. *Voyager 2* crossed the terminal shock boundary several times in one day. This probably occurred because of plasma waves traveling along the shock's turbulent boundary. The position of the terminal shock boundary also changes more slowly with time because of variations in the speed and contents of the solar wind and the galactic wind. It will require numerous measurements in the future to determine the precise geometry of the heliopause.

11. The troposphere is only about 7 km at the poles, while it is about 17 km at the equator. It is greater at the equator because Earth's rotation causes air to move outward, just as water flies out from a spinning sprinkler. Above the troposphere lie, in order, the stratosphere, the mesosphere, and the thermosphere—above which lies the ionosphere.

12. The energy of particles is typically measured in electron volts, eV. One eV is equal to 1.6×10^{-19} Joules, the mks unit most commonly used. Protons in the lower Van Allen belt typically have energies in the 10 to 100 MeV range, where M stands for million.

13. Since the time of Gauss in the mid-nineteenth century, geomagnetists have used various magnetic indices to describe the external magnetic field variations. For example, S_q refers to quiet-day solar variations, and S_d is the daily variation deviating from S_q. The disturbance storm time index, Dst, is based on the average value of the horizontal component of Earth's magnetic field recorded by four observatories located near the equator. Other indices are needed to characterize magnetic storms and substorms, such as the aa index discussed in chapter 6.

14. Innovative methods have been proposed to protect astronauts, such as the construction of mini-magnetospheres around spacecraft. However, problems seem to chase after such innovations (for example, plasma instabilities), and no innovation has thus far proved satisfactory.

15. For readers who have had an undergraduate course in electricity and magnetism, this is referred to as the E × B drift, where E is the electric field and B is the magnetic field.

16. The Lorentz force causes electrons traveling at right angles to a magnetic field to circle around the magnetic field. It is proportional to the strength of the magnetic field and the velocity of the particle. If the electrons also have a velocity component parallel to the magnetic field, the sum of the circular motion and the motion parallel to a magnetic field line is a spiral about the magnetic field line. Positive charges will also spiral around a magnetic field line, but with the opposite circular motion. As an electron approaches a magnetic pole, it encounters a stronger magnetic field. The Lorenz force is then larger, and more energy is portioned into rotational energy. Because energy must be conserved, this requires that the electron have less energy associated with its motion parallel to the magnetic field. Eventually all motion along the magnetic field vanishes at the so-called mirror point of the electron. The electron is then reflected. These concepts are more clearly explained by using some mathematics, as is done in higher-level books, including one of mine: R. T. Merrill, M. W. McElhinny, and P. L. McFadden, *The Magnetic Field of the Earth: Paleomagnetism, the Core, and the Deep Mantle* (San Diego: Academic Press, 1998).

17. Depending on a variety of circumstances, the excited atoms and molecules emit ultraviolet light, infrared light, or X-rays. For example, ionized nitrogen often emits a violet light (around 391 nm) at higher altitudes, but a red light at lower altitudes where more energetic electrons collide with nitrogen.

18. These experiments are discussed in John M. Wallace and Peter V. Hobbs, *Atmospheric Science*, 2nd ed. (San Diego: Academic Press, 2006).

19. As new ice is added to the surface of an ice crystal in a cloud, positive charges (in the form of H_3O^+ ions) diffuse rapidly to the interior of the crystal, leaving the

outer layer negatively charged. A crucial step occurs when the outer layer of an ice crystal with its negative charge is transferred to a larger ice crystal when the two crystals collide. This transfer is related to surface energy. In the interior of a crystal, water molecules are surrounded by other water molecules, but not on the surface. This produces different molecular bonding near the surface relative to the crystal's interior. Because larger grains appear locally flatter at the surface (Earth appears flat to us because of its large radius), there is a difference in bonding of ice molecules at the surface of a large crystal relative to a smaller one. It is easier to add new ice to the larger of two crystals during a collision, and this results in the larger crystal growing at the expense of the smaller one. This also implies that the larger crystal will be negatively charged with respect to the smaller one. Wind within the cloud separates grains with different sizes, resulting in a charge separation. Although the distribution of charges within a single cumulonimbus cloud can be quite complex, on average the top of the cloud is more positively charged than the bottom because the negatively charged ice crystals are heavier.

20. Negatively charged ice crystals build up in what is called the main charging zone, which occurs in that part of the cloud with temperatures between −10°C and −20°C (14°F to −4°F). Positively charged particles dominate the colder part of the cloud that is above the main charging zone. Usually the main charging zone is closer to the bottom of the cloud than the top, but this depends on the temperature distribution within the cloud. Often a weakly positively charged part of the cloud lies below the main charging zone. The distribution of charge within any cloud is difficult to predict because wind within the cloud can redistribute charged particles quickly.

21. This is recorded by a streak camera, which sequentially exposes different lines of pixels. Modern-day streak cameras can record more than 500,000 lines per second.

22. These transient electric phenomena have been given the names of sprites, elves, and blue jets. Sprites are luminous red-orange flashes that extend from the top of a cumulonimbus cloud to about 90 km. They can last up to a few hundred microseconds. A typical elf occurs at a height near 400 km as a dim red light ring with a radius around 60 km. Blue jets are blue-colored lights that extend upward to around 40 to 50 km elevation from the top of a cumulonimbus cloud in the shape of a cone.

Chapter Five

1. Of the five species of salmon in northwestern North America, sockeye salmon are the only one that spends a year in a freshwater lake before heading out to sea.

2. Kirschvink and other scientists find many animals have magnetite particles that produce a signal in the ophthalmic branch of the trigeminal nerve, the largest

nerve in the cranium. This nerve is best known because it is responsible for the facial sensations we experience.

3. Magnetotactic bacteria are found at their highest concentration at, or just below, the oxic-anoxic interface.

4. Magnetotactic bacteria that have greigite magnetosomes live in anaerobic environments. Some other animals have been shown to use maghemite (the cubic form of Fe_2O_3). These crystals also appear to be predominantly in the single-domain state.

5. While about two dozen other great whites have been tagged, data on sharks remains scarce. Some sharks lose their tags or die. A few tagged sharks have been found to migrate from the southern tip of South Africa up to Mozambique and back. Naturally, the behavior of sharks has also been studied elsewhere. Although I am not aware of any record of a shark making as long a voyage as Nicole's, there are enough records of long shark voyages to recognize the difficulty faced in regulating fishery resources: it must be done on a worldwide basis.

6. Some scientists are even investigating whether strong magnetic fields can be used as a shark repellant.

7. Lohmann and his colleagues have also shown that green turtles sense Earth's magnetic field.

8. Let the chromosome carrying the red gene (allele) be labeled X_r, the white one X_w and the male chromosome Y. Then the first female is $(X_r X_r)$ and the first male is $(X_w Y)$. Their offspring (taking one chromosome from each parent) have either $(X_r X_w)$ or $(X_r Y)$. Because the white-eye allele is recessive, the females with $(X_r X_w)$ are red-eyed. However, the next generation (again taking one chromosome from each parent), has some offspring with $(X_w X_w)$, which are white-eyed.

9. Every cell in a human carries DNA that theoretically can be used to make all parts of the body. Yet the body seems to know when and where to make, say, skin cells rather than brain cells. It does this by using segments of DNA that are not genes. Instead these regulatory elements turn genes off or on, depending on the environment. When the temperature of fruit flies increases above 28°C, their life expectancy decreases. During the 1970s, it was discovered that certain sections of non-coding DNA on fruit flies puff up under heat stress. It took nearly 15 years before it was recognized that regulatory elements in other animals also puff up, allowing these elements to be easily identified. In the twenty-first century, so-called cis regulatory elements (*cis* from Latin, meaning "same side of," refers to the elements being on the same strand on DNA) affect various morphological innovations. Although some research done during the twenty-first century suggests that cis regulatory elements appear important to the morphological evolution of fruit flies, debate continues as to how "central" this is in evolution. For an example of this debate, see *Science's News Focus* 321 (August 8, 2008): 760–63.

10. Quantum mechanics provides rules (initially determined from experiments)

on how electron spins are oriented inside atoms. In particular, Wolfgang Pauli (1900–1958) received a Nobel Prize in physics in 1945 for showing that no two electrons could simultaneously occupy the same (quantum) state in an atom. For example, if one electron had its spin pointing "up" in the lowest energy state of an atom, a second electron had to have its spin (magnetic dipole) pointing "down." Often physicists refer to a spin being "up" or "down" (the spins are often drawn as arrows pointing upward or downward), even though in this case "down" does not refer to being in the direction of the gravitational force. In quantum mechanics, the direction of the spins cannot point in arbitrary directions. For example, in the presence of a magnetic field, a spin parallel to the field has lower energy than one in the opposite direction. The rules of quantum mechanics require the spin of two electrons in the ground state (the state in which electrons have a circular orbit closest to the nucleus) be anti-parallel to each other. However, it is possible to have parallel spins in higher-energy states.

11. Some of you may wonder why a pair of radicals is needed. The answer requires more understanding of quantum chemistry than can reasonably be produced in a short note. Thus, the following answer is given for those who already know something about the subject. The radicals under discussion are short-lived and only play an intermediate role in the reaction. Indeed, an individual radical produced by light in cryptochrome is so short-lived it would not play any role in a chemical reaction. Two radicals are required to extend the lifetimes of the radicals. When the spin states of the two radicals are correlated, the radicals can last long enough to affect the rate of a chemical reaction. A correlation can happen when the spin states of the radicals oscillate between the singlet state (in which the electron spins are in opposite directions) and the triplet state (in which a magnetic moment comes from the parallel alignment of the spins). I will not go into how the correlation between spin states occurs, which is not yet known but probably involves so-called hyperfine reactions. A requirement for a chemical reaction to be sensitive to an external magnetic field, such as Earth's, is that either the singlet or the triplet state affects a reaction differently from the other state. Because the magnetic moments of these two states are different, an external magnetic field can then alter the relative longevity of the states sufficiently to affect the rate of a chemical reaction. A reference to one theoretical model is given in note 13.

12. Mora and her colleagues first cut the olfactory nerve used for smell to find that pigeons were still conditioned to a magnetic field. However, when the trigeminal nerve was cut in other pigeons, the pigeons could no longer sense a magnetic field. This is the nerve that Kirschvink and others claim is activated by magnetic particles. See note 2 for elaboration.

13. Birds were the first animals identified as using cryptochrome to sense Earth's magnetic field. T. Ritz, S. Adem, and K Schulten first published the radical-pair mechanism in a 2000 issue of the *Biophysical Journal* with birds in mind. A more

recent review and extension of the theory can be found in Ilia Solov'yov, Danielle Chandler, and Klauss Schulten, "Magnetic Field Effects in *Arabidopsis thaliana* Cryptochrome-1," *Biophysical Journal* 92 (2007): 2711–26.

14. Scientists suggest that a possible explanation for this correlation is that it is beneficial for animals to cooperate when drier ground makes it harder to dig tunnels. In softer ground locations, mole rat species tend to live a solitary existence.

15. Sound waves are rapidly attenuated in the soil. However surface waves (in this case, Rayleigh waves) have longer wavelengths and are not as rapidly attenuated. Surface waves have their largest amplitudes at the surface and their amplitudes rapidly die off with depth. The claim that some mole rats use surface waves rather than sound waves to find mates and termites should be viewed as suggestive, rather than established, science.

16. In 2008 so-called blind mole rats were discovered to be influenced by light, but only barely so. They exhibited a small, but apparently significant, statistical preference against building nests in brightly lit tunnels. However, I am not aware of any scientist who advocates that they sense Earth's magnetic field through photoreceptors, as do birds and fruit flies.

17. As mentioned in chapter 2 and expanded on in the appendix, single-domain magnetite grains are smaller than can be resolved by an optical microscope. In this case, Kirschvink and his colleagues used a variety of tools—including an electron microscope, a superconducting magnetometer, and elemental analysis—to determine that there were ferromagnetic grains ranging in composition from magnetite (Fe_3O_4) to maghemite (cubic Fe_2O_3), which can be formed by the oxidation of magnetite at room temperature, in a variety of brain tissues.

Chapter Six

1. The earliest evidence of life comes from stromatolites, fossils of cyanobacteria. These bacteria have also been called blue-green algae. Although they are not plants, these bacteria, which have their own phylum, still photosynthesize. The chemistry of some sedimentary rocks (such as banded iron formations) also show the evolution from chemically reducing conditions, indicating the lack of atmospheric oxygen, to oxidizing ones.

2. Some readers might be surprised that different mathematical techniques can lead to different outcomes. Many complicated (nonlinear) feedback processes operating in the atmosphere can only be approximately treated, even with today's supercomputers. The type of mathematical approximation used can affect the outcome.

3. Because the variation in Earth's orbit only varies slightly from a circular one over 100,000 years, the amount of heat received from the Sun is by itself insufficient to explain the temperature variation in figure 6.1. Instead some (still unknown)

feedback process must amplify the signal, or the Milankovitch hypothesis is not the correct explanation for the temperature variation.

4. The paper is Arthur Robinson, Noah Robinson, and Willie Soon, "Environmental Effects of Increased Atmospheric Carbon Dioxide," *Journal of American Physicians and Surgeons* 12 (2007): 79–90.

5. The astute reader may recognize that it is not an easy matter to construct a random number. In this case, I constructed the series by first subtracting 3 from pi and then taking successive differences between the resulting numbers.

6. The probability that the contestant's original choice was correct is 1 in 3. If he switches to the other closed door, his chances are 2 in 3. This puzzle is given in many puzzle books and in some elementary books on probability. If you disagree with this answer, try carrying out an experiment to test it. You can use three cards rather than three doors.

7. Geochemical isotopes have been used to identify other times when solar activity was low. When solar activity is low, more high-energy galactic cosmic rays enter our upper atmosphere to produce isotopes of beryllium-10 and carbon-14. For example, when cosmic rays enter the atmosphere, they undergo various transformations leading up to the production of carbon-14. This happens when a neutron collides with nitrogen-14. (A neutron essentially consists of a proton, an electron, and a neutrino.) Minze Stuiver, at the University of Washington, pioneered the use of isotopes of carbon to identify several past intervals of minima—times when solar activity appeared low. In principle, this allows one to test the possibility that solar activity significantly affects climate. Unfortunately, due to a variety of uncertainties, all such tests remain equivocal. More recently, scientists in Germany and Finland have used beryllium-10 in ice cores from Greenland and Antarctica to claim that the Sun has been in a state of unusually high activity for the past 60 years.

8. The average amount of electromagnetic radiation received from the Sun per unit area is called the solar constant. Radiation directed perpendicularly to the outer surface of the atmosphere is measured by aircraft and satellites. The amount of solar radiation received is around 1,366 watts per square meter. Scientists commonly refer to the total radiation from the Sun as the total solar irradiance (TSI). The variation in the solar irradiance from solar minimum to maximum is close to 0.1 percent, or about 1.3 watts per square meter. The solar radiation spectrum is close to that of a black body, an object that absorbs all electromagnetic radiation (including visible light) that falls on it. However, a black body emits radiation that varies depending on its temperature. The Sun emits radiation with wavelengths peaking in the visible range; the amount of radiation is smaller for the adjacent ultraviolet (shorter) and infrared (longer) wavelengths. The Sun is estimated to emit about a billion times more radiation (in all directions) than received by Earth. The radiation received by Earth is affected by many factors, including the local magnetic field on the Sun, as discussed in chapter 4. This means the amount of radiation emitted depends on direction.

9. Courtillot and his colleagues conclude that a variety of evidence links climate variability to solar disturbances. For example, other evidence purported to link geomagnetic secular variation to climate is presented by Courtillot et al.: "Geomagnetic Secular Variation as a Precursor of Climate Change," *Nature* 297 (1982): 386–87.

10. The temperature decrease probably began around 1947 and continued for about 20 years. This decrease occurred while an increase in solar activity was still going on, according to figure 4.2. While this might at first appear to be a serious blow to the theory that the solar magnetic field caused the decline in temperature, it mostly reflects the simplicity of the argument used in this book. Better correlations are claimed when more detailed analyses are performed.

11. Although light specularly reflected at small angles of incidence can be very high, it does not usually reach the viewer, and water is considered to have low albedo. Rather than discussing changes in temperature at Earth's surface, for reasons of clarity of measurements, scientists prefer to use the amount of heat absorbed or radiated at the surface, as measured in watts per meter squared.

12. The burning of coal produces ash containing substantial amounts of magnetite. Magnetite is also a by-product of steel production.

13. Roy Thompson and Frank Oldfield, *Environmental Magnetism* (New York: Allen and Unwin, 1986). Subsequently other books have been published on the subject. For example, Michael Evans and Friedrich Heller, *Environmental Magnetism: Principles and Applications of Enviromagnetics* (San Diego: Academic Press, 2003).

14. The temperature estimates come from several sources, including the use of oxygen-16 and oxygen-18 (two neutrons more than oxygen-16) isotopes in foraminifera, small one-celled marine animals with calcite $(CaCO_3)$ shells. The ratio of oxygen-16 to oxygen-18 used to form the calcite shells of foraminifera is found experimentally to depend on temperature. Paleontologists use this ratio, which remains constant after the animal dies, to estimate paleotemperatures in cores taken from marine sediments.

15. Rodinia is the oldest well-established supercontinent. Although some scientists have suggested the existence of older ones, questions concerning data reliability have led others to more conservative conclusions. For that matter, the precise positions of cratons within Rodinia and the times when this supercontinent first formed and later broke apart are still debated today in the scientific literature.

16. The earliest fossil evidence for dinosaurs comes from rocks from the lowermost Triassic period, which began 250 million years ago. This was long after the 300-million-year-old ice age had ended.

17. Joseph Kirschvink, "Late Protozoic Low-Latitude Global Glaciation: The Snowball Earth," in *The Proterozoic Biosphere: A Multidisciplinary Study*, ed. J. W. Schoff and C. Line (Cambridge: Cambridge University Press, 1992).

18. As described by Gabrielle Walker in *Snowball Earth: The Story of the Great Global Catastrophe that Spawned Life as We Know It* (New York: Three Rivers Press, 2003).

19. Modern plants also slightly prefer carbon-12 to carbon-13.

20. For example, the element iridium is rare at Earth's surface. Although it occurs in some volcanic eruptions, its primary source is from cosmic particles, particularly micro-meteorites. The iridium is hypothesized to have accumulated on ice sheets during Snowball Earth. Upon melting, an iridium layer would be deposited, which has been found at some localities. This evidence was found during the twenty-first century; it was not available to Hoffman and Shrag in 1998.

Epilogue

1. For example, Robert Park, *Voodoo Science: The Road from Foolishness to Fraud* (Oxford: Oxford University Press, 2001).

2. Sabine Begall et al., "Magnetic Alignment in Grazing and Resting Cattle," *Proceedings of the National Academy of Sciences* 105 (2008): 13451–55; Hynek Burda et al., "Extremely Low-Frequency Electromagnetic Fields Disrupt Magnetic Alignment of Ruminants, *Proceedings of the National Academy of Sciences Early Edition*, pnas.0811194106.

3. Many laypersons think of fundamental particles as electrons, protons, and neutrons. Although this can be a useful way to picture matter, physicists now consider only electrons to be fundamental particles, ones that cannot be further subdivided. Both protons and neutrons contain smaller fundamental particles called quarks. For example, a proton contains three quarks, two of which have an electric charge that is two-thirds that of an electron, but of the opposite sign. The third quark has a negative charge that is one-third the size of the electron's. If we sum up the charges (2/3 + 2/3 − 1/3), we find the proton has a positive charge of 1—exactly the same size as, but opposite in sign to, the electron's charge. The electric force between the two quarks with the largest charges is positive and seemingly they should repel each other: the proton would be unstable if only the electromagnetic force were present. The strong nuclear force was proposed to keep these particles from rapidly decaying. The 2004 physics Nobel Prize was awarded to David Gross, David Politzer, and Frank Wilczek, for developing the theoretical constructs underlying this strong nuclear force. The weak nuclear force is associated with radioactive decay.

4. A "general mathematical system" refers to any system broad enough to contain all the formulas of formalized number theory. The mathematically trained reader can find Gödel's proof in numerous places, such as Raymond Wilder's book *Introduction to the Foundation of Mathematics*, 2nd ed. (New York: Wiley, 1956).

5. An example, such as given here, is far from a proof. There may be many examples consistent with a theorem, but all it takes is one counter-example to prove the theorem false. Gödel's theorem was shown to apply to any general mathematical system as defined in the previous note.

Appendix

1. Readers who want a better understanding of rock magnetism should consult the book by David Dunlop and Özden Özdemir, *Rock Magnetism: Fundamentals and Frontiers* (Cambridge: Cambridge University Press, 1997). This is a high-level book intended for scientists.

2. Some readers may be confused as to how this occurs, considering that the magnitude of electron spin moments does not change with temperature. However, the magnitude of temperature fluctuations does increase with temperature (chap. 2), and this breaks down the magnetic order within the grain. The grain's average magnetization decreases because individual atomic moments depart further and further from the average magnetic direction as the temperature is increased.

3. Thermal fluctuations were briefly discussed in chapter 2. When the temperature of a solid is heated, atoms vibrate with increased amplitudes. These vibrations give rise to waves within the solid that are similar to seismic waves in Earth. Because such waves affect the atomic spacing within a crystal, the waves affect the directions of the atomic magnetic moments. The directions of these moments fluctuate about the average direction. The average size of fluctuations increases as the temperature increases.

4. Scientists sometimes use different relaxation constants to define the blocking temperature. Although these definitions usually range from a second to a day, in practice this difference typically changes the estimate of the blocking temperature by a very small amount in single-domain grains.

5. There can be an exchange force (chap. 1, n. 7) that couples two different mineral phases. In such a case, the original direction can be preserved. This possibility has been shown to be true under laboratory conditions when the initial and final grains were single-domain ones. It does not hold for multi-domain grains.

6. There are different methods used to do this, as described in various books including my own: R. T. Merrill, M. W. McElhinny, and P. L. McFadden *The Magnetic Field of the Earth: Paleomagnetism, the Core and the Deep Mantle* (San Diego: Academic Press, 1998). The temperature at which the magnetization is unblocked is not always the same as the temperature at which it was blocked. This is often so for multi-domain grains. Thus, multi-domain grains, which usually carry a soft magnetization, are sometimes found to be unreliable recorders of the paleointensity.

Index